333

Developments in environmental biology of fishes 4

Series Editor
EUGENE K. BALON

Environmental biology of darters

Papers from a symposium on the comparative behavior, ecology, and life histories of darters (Etheostomatini), held during the 62nd annual meeting of the American Society of Ichthyologists and Herpetologists at DeKalb, Illinois, U.S.A., June 14–15, 1982

Edited by
DAVID G. LINDQUIST & LAWRENCE M. PAGE

Reprinted from *Environmental biology of fishes* 11 (2), 1984
with addition of four more papers from the symposium

1984 **DR W. JUNK PUBLISHERS**
a member of the KLUWER ACADEMIC PUBLISHERS GROUP
THE HAGUE / BOSTON / LANCASTER

Distributors

for the United States and Canada:
Kluwer Boston, Inc., 190 Old Derby Street,
Hingham, MA 02043, USA
for the UK and Ireland:
Kluwer Academic Publishers, MTP Press Limited, Falcon House, Queen Square, Lancaster LA1 1RN, England
for all other countries:
Kluwer Academic Publishers Group, Distribution Center,
P.O. Box 322, 3300 AH Dordrecht, The Netherlands

Library of Congress Cataloging in Publication Data

Main entry under title:

Environmental biology of darters.

(Developments in environmental biology of fishes ; 4)
"Reprinted from Environmental biology of fishes 2 (2), 1984 with addition of four more papers from the symposium."
Includes index.
1. Darters (Fishes)--Congresses. I. Lindquist, David G. II. Page, Lawrence M. III. American Society of Ichthyologists and Herpetologists. IV. Series.
QL638.P4E58 1984 597'.58 84-17136

ISBN 90-6193-506-7 (this volume)
ISBN 90-6193-896-1 (series)

Cover design: Max Velthuijs

PRINTED IN THE NETHERLANDS

Contents

Preface, by D.G. Lindquist & L.M. Page 7

An electrophoretic analysis of the *Etheostoma variatum* complex (Percidae: Etheostomatini), with associated zoogeographic considerations, by P.E. McKeown, C.H. Hocutt, R.P. Morgan II & J.H. Howard 9

Ecological and evolutionary consequences of early ontogenies of darters (Etheostomatini), by M.D. Paine 21

Selection of sites for egg deposition and spawning dynamics in the waccamaw darter, by D.G. Lindquist, J.R. Shute, P.W. Shute & L.M. Jones 31

Diets of four sympatric species of *Etheostoma* (Pisces: Percidae) from southern Indiana: interspecific and intraspecific multiple comparisons, by F.D. Martin 37

Life history of the gulf darter, *Etheostoma swaini* (Pisces: Percidae), by D.L. Ruple, R.H. McMichael, Jr. & J.A. Baker 45

Habitat partitioning among five species of darters (Percidae: *Etheostoma*), by M.M. White & N. Aspinwall 55

Life history of the naked sand darter, *Ammocrypta beani,* in southeastern Mississippi, by D.C. Heins & J.R. Rooks 61

Life history of *Etheostoma caeruleum* (Pisces: Percidae) in Bayou Sara, Louisiana and Mississippi, by J.M. Grady & H.L. Bart, Jr. 71

Life history of the bronze darter, *Percina palmaris,* in the Tallapoosa River, Alabama, by W. Wieland 83

Temperature selection and critical thermal maxima of the fantail darter, *Etheostoma flabellare,* and johnny darter, *E. nigrum,* related to habitat and season, by C.G. Ingersoll & D.L. Claussen . . . 95

Morphological correlates of ecological specialization in darters, by L.M. Page & D.L. Swofford . . . 103

A portable camera box for photographing small fishes, by L.M. Page & K.S. Cummings 124

Species and subject index 125

Etheostoma zonale

Etheostoma exile

Percina shumardi

Ammocrypta pellucida

Preface

The following set of papers is mainly a representative sample from 19 presentations at a special symposium on the comparative behavior, ecology, and life histories of darters held during the 62nd (14–15 June, 1982) annual meeting of the American Society of Ichthyologists and Herpetologists on the campus of Northern Illinois University, DeKalb, Illinois, U.S.A.

The idea for a symposium on the behavior and ecology of darters evolved at the April 1981 annual meeting of the Association of Southeastern Biologists in Knoxville, Tennessee (appropriately, the geographic center of darter diversity). One of us (DGL) presided over a serendipitous mini-assemblage of papers on darter behavior, ecology, and natural history and later was encouraged by Robert A. Stiles (Department of Biology, Samford University, Birmingham, AL 35229) to develop a future special symposium on darters. The symposium concept was finalized by R.A. Stiles and D.G. Lindquist at the 3rd biennial conference on Ethology and Behavioral Ecology of Fishes at Illinois State University in Normal, Illinois, in May 1981, when the other of us (LMP) consented to assist both in the soliciting of papers and later in the editing of these proceedings.

The symposium was intended to emphasize the fact that darters offer a superior selection among North American freshwater fishes for purely ethological, ecological and evolutionary investigations, and that studies involving the ethological and ecological approaches to studying darter evolution have been overdue since the pioneering work of Howard Winn in the late 1950's. Darters are relatively sedentary and can be easily observed and experimentally manipulated in their natural habitats. They can be transported alive with relative ease, and readily acclimate to controlled conditions of the laboratory aquarium. The male nuptial color patterns of many of the species are spectacular (equal to many coral reef fishes) and challenge those interested in sexual selection and the functions of color patterns in fishes. The agonistic and reproductive behaviors of darters are complex, poorly understood, and little investigated. Detailed documentation and comparative studies of the evolution of reproductive behaviors in darters beg to be undertaken. Toward this end, one of us (DGL) has submitted a color 16 mm film on the spawning of the waccamaw darter for publication in *Encyclopedia Cinematographica,* and the other (LMP) is attempting to construct a phylogeny of darter reproductive behaviors incorporating as many species as possible. We also want to draw attention to the fact that the restricted distributions of some species make them unfortunate candidates for extinction through man-made or natural habitat perturbations. Preservation and management of darter populations must prevail where changing conditions threaten their habitats.

Authors who contributed papers to the symposium were: Lawrence M. Page and David L. Swofford (keynote address); Frank H. McCormick & Nevin Aspinwall; Michael D. Paine; J. Russell

Rooks & David C. Heins; David G. Lindquist, John R. Shute & Peggy W. Shute; James M. Grady & Kevin S. Cummings; Fred C. Rohde & Steve W. Ross; Patrick E. O'Neil; Larry A. Greenberg; Robert A. Stiles; Matthew M. White & Nevin Aspinwall; William J. Matthews & Jeffrey R. Bek; Noel M. Burkhead, Daphne Ebaugh & Robert E. Jenkins; William L. Pflieger; Werner Wieland; F. Douglas Martin; James M. Grady & Henry L. Bart; Robert A. Stiles & Richard T. Bryant; Paul E. McKeown, Charles Hocutt, James Howard & Raymond P. Morgan II. Special gratitude goes to William J. Matthews, University of Oklahoma Biological Station and Robert A. Stiles for chairing symposium sessions; David W. Greenfield, Northern Illinois University (and his assistants) for graciously accepting and coordinating the symposium as host for the ASIH meeting; Eugene K. Balon for patient advice on editorial procedures; and the referees for courteous and timely reviews of the manuscripts (their names will be listed in the final issue of the journal volume in which these papers appear).

All authors who presented papers at the symposium were asked to submit, at their discretion, manuscripts to be considered for this special issue. The contents of these proceedings were scrutinized via the normal external review process. Papers presented at the symposium and in the present issue include studies on life histories, habitat selection, zoogeography, and morphological evolution in relation to environment. Although life-history studies sometimes are criticized as being too descriptive and superficial to address interesting questions, they often provide the empirical basis from which interesting questions arise. Once the ecological diversity of a group of organisms is understood, more specific questions can be meaningfully addressed. The first paper examines geographic variation in five closely related species of *Etheostoma* and the zoogeographic implication of the variation. The next paper examines the significance of the early ontogeny. The study on selection of sites for egg deposition in *E. perlongum* expands on the earlier published study on the life history of this species. Two papers address habitat partitioning: one examines diet diversification among four coinhabiting darters, and one examines physical characteristics of the environments of five coinhabiting species. The collection of papers continues with life history studies on *Ammocrypta beani, Percina palmaris, Etheostoma caeruleum, E. swaini,* and a study on temperature selection, the last two papers being not a part of the proceedings but submitted to the Journal directly. The last paper, based on the keynote address, attempts to synthesize the relationships among morphology, environment and evolution.

A symposium should serve to illustrate what needs to be done. Clearly, much pertaining to the ecology and behavior of darters remains to be done. More life histories need to be elucidated, especially those of highly divergent species such as *A. asprella, E. cinereum,* and *E. tuscumbia.* Questions begging to be addressed range from those appearing to be simple (Where and how do species of *Ammocrypta* spawn?), to those which obviously are complex (Why are there so many species in the subgenus *Nothonotus* (ca. 15) and so few species of *Allohistium* (1), *Litocara* (2), and *Boleosoma* (5)?). How do syntopic species of darters partition habitats? What are the habitat requirements of darter larvae? What environmental variables initiate, and terminate, the spawning season? Do darters migrate and, if so, for what reasons? Does a female lay one set of mature eggs and then during the same spawning season develop another set? Do females select among potential breeding males, and if so, is it on the basis of morphological, behavioral, or ecological (e.g. territory quality) characteristics? Do males select among females?

It is our sincere hope that this small group of papers will stimulate further research on the behavior and ecology of darters. Future studies should attempt to integrate and synthesize the growing body of pertinent literature with the aim of elucidating tactics and strategies that have allowed darters to become such successful exploiters of North American freshwater environments.

David G. Lindquist & Lawrence M. Page

An electrophoretic analysis of the *Etheostoma variatum* complex (Percidae: Etheostomatini), with associated zoogeographic considerations

Paul E. McKeown[1], Charles H. Hocutt[2], Raymond P. Morgan II[1] & James H. Howard[3]
[1] *University of Maryland, Center for Environmental and Estuarine Studies, Appalachian Environmental Laboratory, Frostburg State College Campus, Gunter Hall, Frostburg, MD 21532, U.S.A.*
[2] *University of Maryland, Center for Environmental and Estuarine Studies, Horn Point Environmental Laboratories, Box 775, Cambridge, MD 21613, U.S.A.*
[3] *Biology Department, Frostburg State College, Frostburg, MD 21532, U.S.A.*

Keywords: Faunal distribution, Genic variation, Pleistocene glaciation, Relict populations

Synopsis

The *Etheostoma variatum* complex is comprised of five species (*E. euzonum, E. kanawhae, E. osburni, E. tetrazonum, E. variatum*) distributed from the Allegheny River, New York, to the White River, Arkansas. Electrophoretic data provide evidence of a division of the complex into two geographic units: *E. variatum, E. kanawhae,* and *E. osburni* in the Appalachian region, and *E. euzonum* and *E. tetrazonum* in the Ozarks. Genic variation exists also between the Sac and Big river populations of *E. tetrazonum.* Genic variation and present faunal distributions suggest that an ancestral stock was widely distributed in Teays and Old Mississippi rivers but separated by a Pleistocene ice advance. Some populations survived in an Ozarkian refugium, while more eastern populations, such as the precursor to *E. variatum,* may have evolved in a southern refuge of the developing Ohio River. The Teays (New) River gorge, including Kanawha Falls, has prevented *E. variatum* from invading territory occupied by *E. osburni* and *E. kanawhae.*

Introduction

The *Etheostoma variatum* complex is a group of five closely related largely allopatric species ranging from the Allegheny River drainage of New York and Pennsylvania southwest to the White River system of Arkansas and Missouri (Fig. 1). The complex encompasses five recognized species, one of which includes two subspecies: *E. variatum, E. osburni, E. kanawhae, E. tetrazonum, E. euzonum euzonum* and *E. e. erizonum* (Hubbs & Black 1940, Raney 1941).

Present address: PE McK, Muddy Run Ecological Laboratory, P.O. Box 10, Drumore, PA 17518, U.S.A.
CHH, Department of Ichthyology and Fisheries Science, J.L.B. Smith Institute for Ichthyology, Rhodes University, Grahamstown, South Africa.

Etheostoma variatum, the variegate darter, is confined to the Ohio River drainage from south-central Indiana upstream to the headwaters of the Allegheny and Monongahela system (Gilbert 1980) excluding the upper Kanawha (New) River system of West Virginia, Virginia, and North Carolina. Zoogeographically, that portion of the Kanawha River system above Kanawha Falls is synonymous with the New River (Hocutt 1979). *Etheostoma osburni,* the finescale saddled darter, is endemic to the New River (Hubbs & Trautman 1932, Hocutt et al. 1980a). Although it is a locally common endemic in the lower half of the New River drainage in West Virginia and Virginia, there are no records in the New River system upstream from Reed Creek, Virginia, where it is replaced by *E. kanawhae. Etheostoma kanawhae,* the Kanawha darter, is endemic to the upper New River and its tributaries in

David G. Lindquist & Lawrence M. Page (ed.), Environmental biology of darters. ISBN 90-6193-506-7

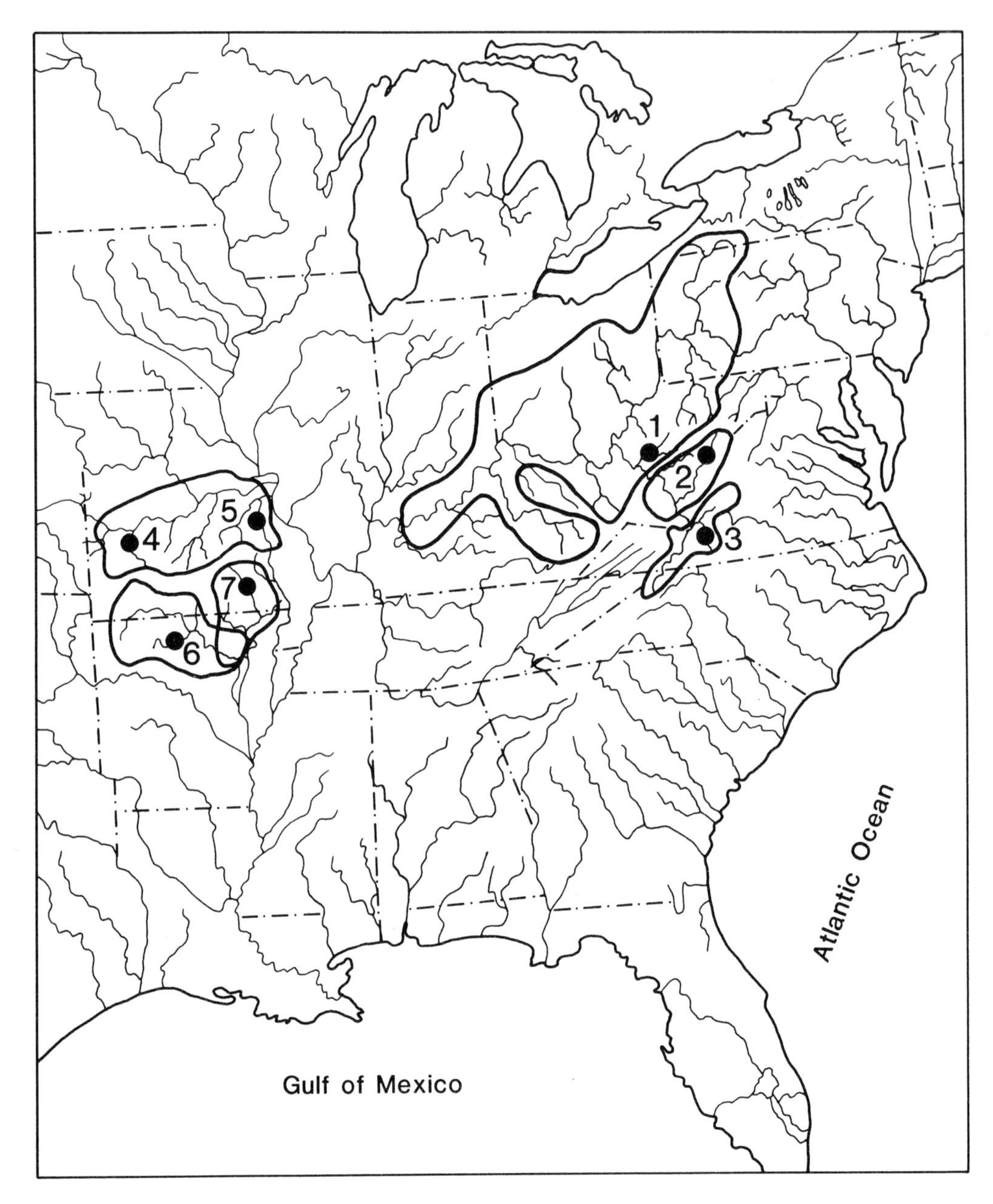

Fig. 1. Geographic distribution of (1) *E. variatum,* (2) *E. osburni,* (3) *E. kanawhae* (4) *E. tetrazonum* (Osage River system), (5) *E. tetrazonum* (Meramec River system), (6) *E. euzonum euzonum,* and (7) *E. euzonum erizonum.* Collection sites designated by ●.

North Carolina and Virginia (Raney 1941), and is possibly limited in distribution by water hardness (Ross & Perkins 1959) or competition with *E. osburni* (Hocutt et al. 1980b).

Etheostoma tetrazonum, the Missouri saddled darter, is endemic to Missouri and restricted to streams draining north off the Ozark Plateau (Hubbs & Black 1940). *Etheostoma euzonum,* the Arkansas saddled darter, replaces the closely related *E. tetrazonum* in the southern Ozarks, being endemic to the White River drainage in the Ozark mountains of Arkansas and Missouri. Two subspecies have been described, *E. e. euzonum* and *E. e. erizonum* (Hubbs & Black, 1940). *Etheostoma e.*

euzonum occurs in the White River system proper above Batesville, Arkansas, and *E. e. erizonum* is known from the Black River (Stauffer et al. 1980).

Although morphometric analysis of this complex has been reported (Pflieger 1971, Page 1981), no research has investigated biochemical affinities. The purpose of this study was to utilize electrophoretic techniques in assessing the phylogeny and zoogeographic history of the *E. variatum* complex.

Materials and methods

Twenty-five specimens were collected from each population and either immediately frozen in dry ice or maintained alive in an aerated tank while being transported to the Appalachian Environmental Laboratory. Specimens of *E. variatum* were collected from Elk River, Webster County, West Virginia; of *Etheostoma osburni* from East Fork Greenbrier River, Pocahontas County, West Virginia; and of *E. kanawhae* from West Fork Little River, Floyd County, Virginia (Fig. 1).

Two populations of *E. tetrazonum* were sampled. An eastern population was sampled from Big River (Meramec River system) at Route H Bridge, Jefferson County, Missouri; it will hereafter be referred to as *E. tetrazonum* (e.). A western population (w.) was sampled from Sac River (Osage River system), St. Claire County, Missouri. *Etheostoma e. euzonum* was collected from Buffalo River off State Route 14, Marion County, Arkansas, and *E. e. erizonum* from Current River off State Route 142, Ripley County, Missouri (Fig. 1). Specimens not used for experimentation are preserved at the Appalachian Environmental Laboratory.

Brain and muscle tissues were electrophoretically analyzed (Table 1). Horizontal starch gel electrophoresis followed procedures outlined in Selander et al. (1971) and May (1975). Genetic distances (Nei 1972, Wright 1978) and similarities (Nei 1972) were determined for all populations (Table 2). From the matrix of Prevosti distance coefficients, an unrooted tree was constructed using the distance Wagner procedure of Farris (1972) as modified by Swofford (1981). The network was then rooted at the midpoint of the largest path separating a pair of populations (Farris 1972).

Chi-square goodness-of-fit analysis, used to compare the expected and observed allelic frequencies as related to Hardy-Weinberg proportions was employed at each locus for each population. Nei's (1973) component analysis of genetic differentiation was employed as a method of examining genetic diversity within and among populations. To determine the ability of the cladogram to reflect the distance matrix, a number of statistical indices were considered; Sneath & Sokal's (1973) cophenetic correlation coefficient; Farris' (1972) F value, Prager & Wilson's (1976) F value and the percent standard deviation of Fitch & Margoliash (1967).

Results

Specimens were analyzed for gene products of 27 loci; a total of 43 presumptive alleles were resolved. Nine of these 27 loci proved to be polymorphic, while the remaining 18 were monomorphic (Table 1). Banding patterns for polymorphic loci which could be consistently resolved are discussed below (Fig. 2, 3):

Esterase: Three loci were present for the esterase enzyme. At Est-1, three alleles were resolved. *Etheostoma variatum, E. osburni, E. tetrazonum* (e.), *E. e. euzonum* and *E. e. erizonum* were all fixed for the slow variant 'BB'. Both alleles 'A' and 'B' were resolved for *E. kanawhae. E. tetrazonum* (w.) was fixed for the fast variant 'CC' (Fig. 2). Est-2 could not be consistently scored for all populations and was not considered. Est-3 was monomorphic for all populations.

Glutamate-oxaloacetate transaminase: Two loci were resolved for this enzyme. GOT-1 was monomorphic over all populations. GOT-2 possessed two alleles. *Etheostoma variatum* was homozygous for the 'AA' slow variant while all other populations were homozygous for the 'BB' fast variant (Fig. 2).

Isocitrate dehydrogenase: Three alleles were present for this isozyme at one locus. *Etheostoma variatum* and *E. kanawhae* were limited to alleles

Table 1. Allele frequencies in representative samples of populations of the *E. variatum* complex. (1) *E. variatum*, (2) *E. kanawhae*, (3) *E. osburni*, (4) *E. tetrazonum* (Meramec River system), (5) *E. tetrazonum* (Osage River system), (6) *E. euzonum euzonum*, (7) *E. euzonum erizonum*.

Loci	Buffer	Populations						
		(1)	(2)	(3)	(4)	(5)	(6)	(7)
ADH	IV	A = 1.00	1.00	1.00	1.00	1.00	1.00	1.00
ALD	II	A = 1.00	1.00	1.00	1.00	1.00	1.00	1.00
CK-1	I	A = 1.00	1.00	1.00	1.00	1.00	1.00	1.00
CK-2	I	A = 1.00	1.00	1.00	1.00	1.00	1.00	1.00
EST-1	I	A = 0.00	0.1957	0.00	0.00	0.00	0.00	0.00
		B = 1.00	0.8043	1.00	1.00	0.00	1.00	1.00
		C = 0.00	0.00	0.00	0.00	1.00	0.00	0.00
EST-3	I	A = 1.00	1.00	1.00	1.00	1.00	1.00	1.00
FUM	I	A = 1.00	1.00	1.00	1.00	1.00	1.00	1.00
GAPDH	II	A = 1.00	1.00	1.00	1.00	1.00	1.00	1.00
GDH	IV	A = 1.00	1.00	1.00	1.00	1.00	1.00	1.00
GOT-1	III	A = 1.00	1.00	1.00	1.00	1.00	1.00	1.00
GOT-2	III	A = 1.00	0.00	0.00	0.00	0.00	0.00	0.00
		B = 0.00	1.00	1.00	1.00	1.00	1.00	1.00
IDH	II	A = 0.00	0.00	0.00	0.00	0.00	0.00	0.1800
		B = 0.9200	0.9400	1.00	1.00	1.00	1.00	0.8200
		C = 0.0800	0.0600	0.00	0.00	0.00	0.00	0.00
LDH-1	II	A = 1.00	1.00	1.00	1.00	1.00	1.00	1.00
LDH-2	II	A = 0.5600	1.00	1.00	1.00	1.00	1.00	1.00
		B = 0.4400	0.00	0.00	0.00	0.00	0.00	0.00
MDH-1	III	A = 1.00	1.00	1.00	1.00	1.00	1.00	1.00
MDH-2	III	A = 1.00	1.00	0.00	1.00	1.00	1.00	1.00
		B = 0.00	0.00	1.00	0.00	0.00	0.00	0.00
MUS-1	I	A = 1.00	1.00	1.00	1.00	1.00	1.00	1.00
MUS-2	I	A = 1.00	1.00	1.00	1.00	1.00	1.00	1.00
MUS-3	I	A = 1.00	1.00	1.00	1.00	1.00	1.00	1.00
6-PGDH	I	A = 1.00	1.00	1.00	1.00	1.00	1.00	1.00
PGI-1	I	A = 0.00	0.3400	0.00	0.1333	0.0667	0.00	0.00
		B = 1.00	0.660	1.00	0.8667	0.8667	0.9800	0.9600
		C = 0.00	0.00	0.00	0.00	0.0667	0.0200	0.0400
PGI-2	I	A = 1.00	1.00	1.00	0.00	0.00	0.00	0.00
		B = 0.00	0.00	0.00	1.00	0.9333	0.8400	0.9800
		C = 0.00	0.00	0.00	0.00	0.0667	0.1600	0.0200
PGM	I	A = 0.0600	0.00	0.00	0.00	1.00	0.00	0.00
		B = 0.0400	0.00	0.00	0.00	0.00	0.00	0.00
		C = 0.900	1.00	1.00	0.9667	0.00	1.00	1.00
		D = 0.00	0.00	0.00	0.0333	0.00	0.00	0.00
PMI-1	II	A = 1.00	1.00	1.00	1.00	1.00	1.00	1.00
PMT-2	II	A = 1.00	1.00	1.00	1.00	1.00	1.00	1.00
SODH	III	A = 1.00	1.00	1.00	1.00	1.00	1.00	1.00
XDH	III	A = 0.00	0.0400	0.00	0.00	0.0333	0.00	0.00
		B = 1.00	0.9600	1.00	1.00	0.8000	1.00	1.00
		C = 0.00	0.00	0.00	0.00	1.667	0.00	0.00

Table 2. Prevosti genetic distance (above the diagonal), Nei's genetic distance (below the diagonal), and Nei's genetic identity (in parenthesis).

	Populations						
	(1)	(2)	(3)	(4)	(5)	(6)	(7)
(1) *E. variatum*		.079	.097	.102	.178	.098	.102
(2) *E. kanawhae*	.053 (.948)		.061	.157	.130	.077	.065
(3) *E. osburni*	.087 (.917)	.045 (.956)		.080	.160	.075	.082
(4) *E. tetrazonum* (e.) Meramec River System	.088 (.916)	.042 (.959)	.078 (.925)		.086	.012	.014
(5) *E. tetrazonum* (w.) Osage River System	.167 (.846)	.116 (.890)	.161 (.851)	.078 (.924)		.089	.093
(6) *E. euzonum euzonum*	.081 (.922)	.040 (.961)	.071 (.931)	.002 (.998)	.080 (.923)		.013
(7) *E. euzonum erizonum*	.088 (.916)	.045 (.056)	.078 (.925)	.002 (.998)	.081 (.924)	.002 (.998)	

'B' and 'C'. *Etheostoma e. erizonum* was limited to alleles 'A' and 'B'. *Etheostoma osburni, E. tetrazonum* (e.) and *E. tetrazonum* (w.) and *E. e. euzonum* were all monomorphic for allele 'BB' (Fig. 2).

Lactate dehydrogenase: Two loci were resolved for this enzyme. LDH-1 was monomorphic for all populations. Two alleles, designated 'A' and 'B', were present at LDH-2 for *E. variatum*. All other populations were monomorphic for allele 'A' (Fig. 2).

Malate dehydrogenase: A four-banded electromorph was resolved for all populations, coded for at two presumptive loci. All populations were monomorphic at MDH-1. At MDH-2, allele differentation was based on the relative migration distance of the second band. This second band consistently migrated a shorter distance from the origin for the *E. osburni* population than in all other populations considered, and was designated allele 'B'. *Etheostoma osburni* was fixed for allele 'B' while all other populations were fixed for allele 'A' (Fig. 3).

Phosphoglucose isomerase: PGI was determined to be a dimer coded for at two loci. At PGI-1, *E. variatum* and *E. osburni* were fixed for allele 'B'. In *E. kanawhae* and *E. tetrazonum* (e.), both alleles 'A' and 'B' were present. *Etheostoma e. euzonum* and *E. e. erizonum* possessed alleles 'B' and 'C'. *Etheostoma tetrazonum* (w.) possessed alleles 'A', 'B', and 'C'. At PGI-2, *E. kanawhae, E. variatum,* and *E. osburni* were all fixed for allele 'A'. *Ethe-*

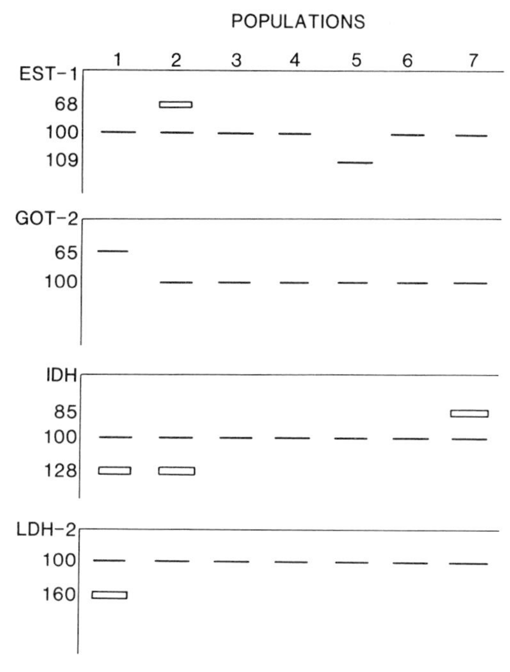

Fig. 2. Banding patterns and relative migration distances of polymorphic loci for (1) *Etheostoma variatum,* (2) *E. kanawhae,* (3) *E. osburni,* (4) *E. tetrazonum* (e.), (5) *E. tetrazonum* (w.), (6) *E. euzonum euzonum,* and (7) *E. euzonum erizonum.*

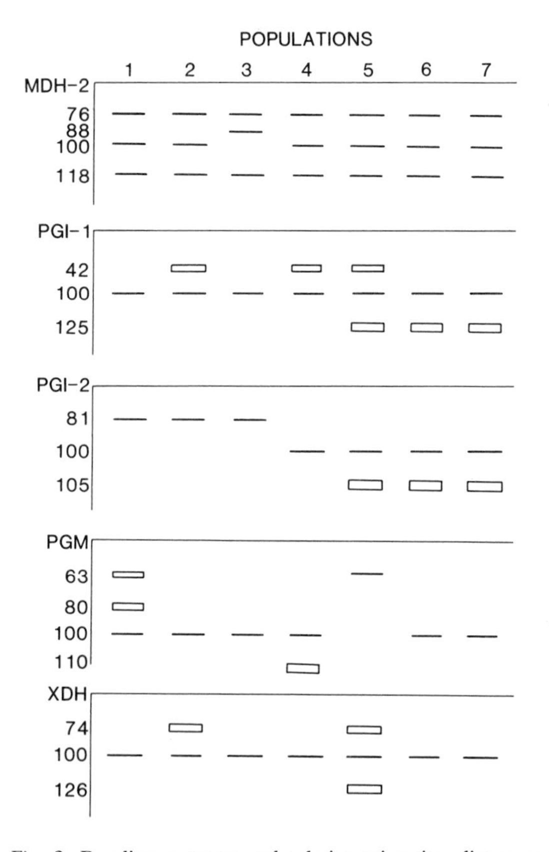

Fig. 3. Banding patterns and relative migration distances of polymorphic loci for (1) *Etheostoma variatum,* (2) *E. kanawhae,* (3) *E. osburni,* (4) *E. tetrazonum* (e.), (5) *E. tetrazonum* (w.), (6) *E. euzonum euzonum,* and (7) *E. euzonum erizonum.*

ostoma tetrazonum (e.) was fixed at allele 'B'. Both alleles 'B' and 'C' were present for *E. e. euzonum, E. e. erizonum,* and *E. e. tetrazonum* (w.) (Fig. 3).

Phosphoglucomutase: Encoded at a single presumptive locus, four alleles were present over all populations examined. *Etheostoma variatum* possessed three of these alleles, designated 'A', 'B', and 'C'. *Etheostoma kanawhae, E. osburni, E. e. euzonum,* and *E. e. erizonum* were fixed at allele 'C'. *Etheostoma tetrazonum* (e.) were polymorphic for alleles 'C' and 'D', while *E. tetrazonum* (w.) was monomorphic for allele 'A' (Fig. 3).

Xanthine dehydrogenase: Three alleles were present at this single-locus system. *Etheostoma tetrazonum* (w.) possessed all three alleles. Alleles 'A' and 'B' were present in *E. kanawhae.* All other populations were monomorphic for allele 'B' (Fig. 3).

The Prevosti genetic distance coefficient (Wright 1978) ranged from a low of 0.012 between *E. e. euzonum* and *E. tetrazonum* (e.) to a high of 0.178 between *E. variatum* and *E. tetrazonum* (w.). Nei's (1972) genetic distance coefficient ranged from a low of 0.002 between *E. e. euzonum* and *E. tetrazonum* (e.), *E. e. erizonum* and *E. tetrazonum* (e.), and *E. e. euzonum* and *E. e. erizonum* to a high of 0.167 between *E. variatum* and *E. tetrazonum* (w.) (Table 2).

Chi-square goodness-of-fit analysis applied at 17 individual loci resulted in two significant deviations at the 0.05 level; XDH in the *E. kanawhae* population ($X^2 = 30.00$, df = 1) and PGM in *E. tetrazonum* (e.) ($X^2 = 30.00$, df = 1).

Cophenetic analysis of genetic differentiation (Nei 1973) was employed using both a single and double cluster. In the single cluster, all populations included, the gene differentiation relative to the total population ($\bar{G}_{ST}$) was 0.751. The gene diversity in the total population ($\bar{H}_T$) was 0.076, and the gene diversity between populations ($\bar{D}_{ST}$) was 0.057. Using the double cluster method, the populations were separated into two groups. The first, consisting of *E. variatum, E. kanawhae,* and *E. osburni,* produced a $\bar{G}_{ST} = 0.639$, $\bar{H}_T = 0.061$, and a $\bar{D}_{ST} = 0.039$. The second, made up of *E. e. euzonum, E. e. erizonum, E. tetrazonum* (e.), and *E. tetrazonum* (w.) produced a $\bar{G}_{ST} = 0.638$, $\bar{H}_T = 0.045$, and a $\bar{D}_{ST} = 0.029$.

The distance Wagner tree obtained using Prevosti genetic distance coefficients (Wright 1978) is shown in Figure 4. A relatively close fit of the computed path length distance to the observed distance is indicated by the following goodness-of-fit statistics: cophenetic correlation coefficient (Sneath & Sokal 1973) = 0.997, Farris (1972) *F* value = 0.036, Prager & Wilson (1976) *F* value = 2.085, percent standard deviation (Fitch & Margoliash 1967) = 4.229.

Discussion

Chi-square goodness-of-fit analysis was used to compare observed and expected allele frequencies as related to Hardy-Weinberg proportions. Only

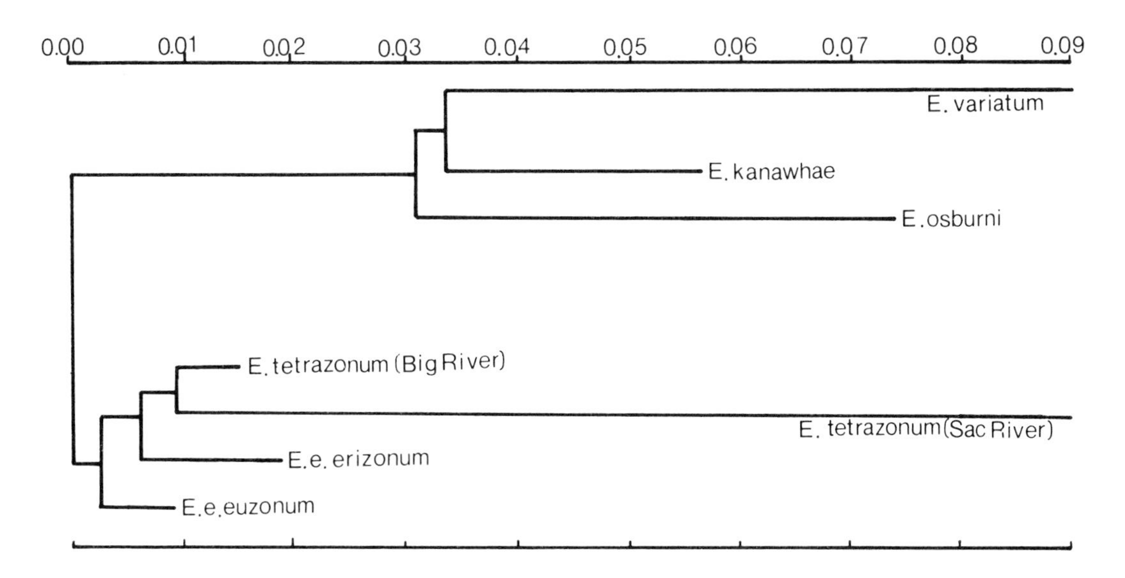

Fig. 4. Distance Wagner tree obtained using Prevosti distances.

two of 17 Chi-square tests deviated significantly from Hardy-Weinberg proportions. This deviation may be caused by scoring ambiguity; non-detected protein differences as nucleotide changes may occur without altering the net charge of the polypeptides (Avise 1974).

Four loci were considered important in defining relationships in the *E. variatum* complex: IDH, PGI-1, PGI-2, and PGM (Fig. 2, Fig. 3). Other loci which were examined, EST-1, GOT-2, LDH-2, MDH-2, and XDH, revealed variation due to alleles unique to single populations. *Etheostoma variatum* was distinguished by the fixation of GOT-2^A, and the presence of LDH-2^B and PGM^B. *Etheostoma kanawhae* possessed the unique allele EST-1^A, and *E. osburni* was fixed for MDH-2^B. *Etheostoma tetrazonum* (e.) possessed a unique allele PGM^D while *E. tetrazonum* (w.) was distinguished by the fixation of EST-1^C and the presence of XDH^C. *Etheostoma e. erizonum* possessed unique allele IDH^A (Fig. 2, 3).

Component analysis of genetic differentiation (Nei 1973) indicates a higher degree of variation between populations ($\bar{D}_{ST} = 0.057$) than within populations ($\bar{H}_{ST} = 0.019$) when all populations examined were included in one cluster. These results are to be expected since each population represented a species, with the exceptions of the two populations of *E. tetrazonum* and *E. euzonum.* A substantial decrease in the average diversity between populations ($\bar{D}_{ST}$) was observed when a two-cluster technique was employed. One cluster was composed of the eastern species: *E. variatum, E. osburni,* and *E. kanawhae,* and the second cluster was composed of the Ozarkian populations: *E. tetrazonum* (e.), *E. tetrazonum* (w.), *E. e. euzonum,* and *E. e. erizonum.* There was no corresponding change in variation within each population ($\bar{H}_{ST}$), thus implying that a substantial part of the variation measured between populations in the single cluster technique was due to variation between the eastern and western components of the *E. variatum* complex.

The Wagner procedure (Farris 1972) was selected to represent the data matrix since it makes no assumptions concerning uniform evolutionary rates and does not require information on whether alleles are primitive or ancestral in constructing an unrooted tree. The midpoint method of Farris (1972) which involves a fairly weak assumption regarding rate uniformity was also employed.

The distance Wagner tree revealed an early divergence of two groups within the *E. variatum* complex; *E. variatum, E. kanawhae,* and *E. osburni* (presently occupying the Appalachian region) and *E. tetrazonum* and *E. euzonum* (presently occupying the Ozarkian region). Within the Appalachian group, *E. variatum* aligned more closely to *E. kanawhae* than to *E. osburni*, while in the Ozarkian group intraspecific populations aligned as expected. A surprising result of the Wagner network was the variation observed between the Big River (e.) and Sac River (w.) populations of *E. tetrazonum* ($D_P = 0.086$), and the apparent similarity between populations of *E. tetrazonum* (e.) and *E. euzonum* ($D_P = 0.012$, $D_P = 0.014$, respectively) (Table 2, Fig. 4).

Variation observed between geographically separate populations may be a result of varying rates of evolutionary change associated with genetic drift and differential selection. Apparent similarity between the eastern population of *E. tetrazonum* and *E. euzonum* may be due to a slower rate of evolutionary change for these populations undergoing similar selective pressures. Conversely, variation observed in the western population of *E. tetrazonum* may be due to rapid evolutionary change coupled with intensive differential selection.

Non-uniform evolutionary rate is important when considering species that may have been subject to rapid evolutionary change in exploiting diverse ecological opportunities (Wright 1982), opportunities which may have been available to the *E. variatum* complex during Pleistocene glaciation. The *E. variatum* complex may once have enjoyed a widespread preglacial distribution from the upper Teays system westward to the lower Missouri system (Pflieger 1971). The advance of glacial ice and corresponding climatic changes lead to rearrangements of distributional patterns of fishes (Cross 1970, Pflieger 1971, Hocutt et al. 1978, Hocutt 1979). Some northern populations likely migrated southward, many were eliminated, and others were restricted to relatively small areas. Isolated relict populations apparently occurred as a result of two events of Pleistocene history, one during the height of the glaciations when populations were restricted to small refugia, and the second when post-glacial lakes and streams began to recede (Briggs 1983).

The ancestral stock of the *E. variatum* complex probably became localized in the Ozark Uplands and unglaciated portions of the upper Ohio Valley during an early ice advance, giving rise to geminate forms after failing to reoccupy much of their former range (Pflieger 1971). Members of the complex endemic to the Ozark Uplands are *E. e. euzonum*, *E. e. erizonum*, and *E. tetrazonum*. Pflieger (1971) attributes the differentiation seen in *E. euzonum* to the Coastal Plain being an effective barrier to gene exchange for upland species (Jenkins et al. 1972, Cashner & Suttkus 1977). Differentiation observed in the Meramec (e.) and Osage river (w.) populations of *E. tetrazonum* may be attributed to the lowland, big river habitat of the Missouri and Mississippi rivers that serves as an effective barrier to dispersal for species adapted to clear upland streams (Fig. 1). For instance, the distribution of *Percina cymatotaenia* in Missouri (Osage River system) and the closely related *Percina* (*Odontopholis*) sp. in Kentucky parallels the scenario described here for the *E. variatum* complex, with the disruption of a once wider distribution by glacial events, followed by isolation and speciation in southern refugia.

Etheostoma variatum is widely distributed in tributaries of the Ohio River, although not found above Kanawha Falls and not penetrating far into glacial regions (Fig. 1). The inception of Kanawha Falls, long considered a barrier to fish dispersal in the Upper Kanawha (New) River drainage, is unknown (Lachner & Jenkins 1971). Miller (1968) believed it was probably cut during the first major glacial recession, but lacked supportive data. The falls, in concert with the cataracts of the New River gorge, act as a faunal filter with perhaps none of them being completely effective within itself, but being important collectively (Hocutt 1979) (Fig. 5).

Although various proposals can be offered for the evolution of *E. variatum*, *E. osburni*, and *E. kanawhae*, none is without its limitations. The distribution of *E. osburni* is perplexing in that it occupies a virtual parapatric distribution between two closely allied species. The following account is offered for the formation of the eastern component of the *E. variatum* complex, however, with the qualifications mentioned above.

It may well be that the inception of Kanawha Falls isolated three components of the ancestral

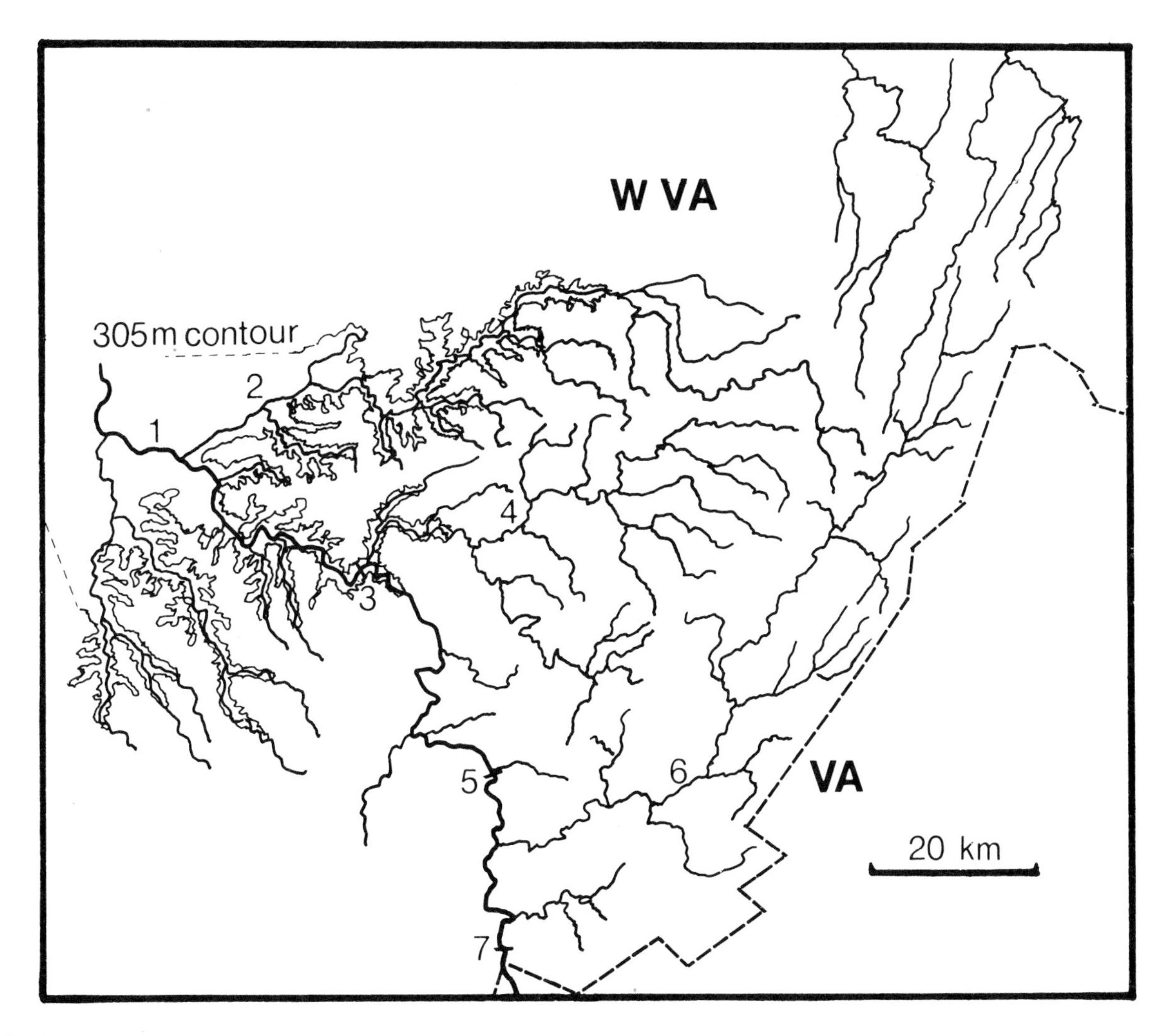

Fig. 5. Map depicting New River system in West Virginia with associated barriers to fish dispersal. River systems and barriers depicted are (1) New River, (2) Elk River, (3) Kanawha Falls, (4) Gauley River, (5) Sandstone Falls, (6) Greenbrier River, and (7) Wylie Falls. Indicated 305 m contour is reflective of the presumed level of Teays Lake, or Lake Tright, a Pleistocene glacial lake (modified from Hocutt 1979).

stock into (a) lower Teays River and its tributaries; (b) Teays River above Kanawha Falls, excluding Gauley River; and (c) Old Gauley River. With the advance of the Nebraskan and Kansas events Lake Tight or Teays Lake was formed (Hocutt 1979), thus limiting the lower Teays stock to southern tributaries which served as refugia. In such a manner *E. variatum* evolved and expanded its range during interglacial and post-glacial times. Kanawha Falls in concert with the New River george prevented upstream dispersal to the Upper Teays. Isolation and subsequent speciation ultimately led to the formation of *E. kanawhae* in the upper Teays and *E. osburni* in the Old Gauley River. Dismemberment of Old Gauley River and formation of Greenbrier River (Hocutt 1979) would have effectively introduced *E. osburni* into the Upper Teays (New) River system (Fig. 5).

Two sources of biological evidence are offered to support this hypothesis: (1) *E. kanawhae* replaces *E. osburni* only a few kilometers upstream of the confluence of New and Greenbrier rivers and (2) *E. caeruleum* is another species distributed throughout Gauley River (Hocutt et al. 1979) that has a localized distribution in New River not far upstream from the mouth of Greenbrier River. Once in New River, *E. osburni* could have easily dispersed downstream through the gorge. Upstream dispersal was hampered by changes in water quality and physiographic features identified with the New River Valley as it passes downstream from the Blue Ridge and Ridge and Valley Provinces to the Appalachian Plateau (Ross & Perkins 1959) and putative competition with *E. kanawhae.* It should be noted that the transition area between the Ridge and Valley and Appalachian Plateau provinces is virtually at the upstream limit of *E. osburni* (Hocutt et al. 1980a) and downstream limit of *E. kanawhae* (Hocutt et al. 1980b).

In summary, biochemical data provide evidence of relationships consistent with the geographic distributions of species comprising the *E. variatum* complex. Genic variation and existing faunal distributions concurrently suggest that a large-scale speciation event, Pleistocene glaciation, has played a major role in the isolation and subsequent speciation of these fishes.

Isolation has resulted in distinct genic variation between the Sac and Big river populations of *E. tetrazonum.* Unfortunately, no morphometric or meristic research has examined specimens from both river systems. If future morphometric analysis of specimens from both river systems support the biochemical data, then a re-evaluation of the taxonomic status of these populations is in order.

Acknowledgements

Assistance in collections was provided by R.F. Denoncourt, E.F. Esmond, W.L. Goodfellow, R.E. Smith, and J.R. Stauffer Jr. Margaret A. Eckman, J.M. McKeown, and B.J. Tollinger aided in preparation of this manuscript. Frank B. Cross, L.M. Page, and W.L. Pflieger contributed helpful comments on fish distribution and zoogeography. David L. Swofford provided a valuable critique of the data analysis. This is Contribution No. 1415-AEL of the Appalachian Environmental Laboratory, University of Maryland, Center for Environmental and Estuarine Studies.

References cited

Avise, J.C. 1974. Systematic value of electrophoretic data. Syst. Zool. 23: 1–4.

Briggs, J.C. 1983. Introduction to the zoogeography of North American freshwater fishes. *In*: C.H. Hocutt (ed.) Zoogeography of North American Freshwater Fishes, J. Wiley & Sons, New York (in press).

Cashner, R.C. & R.D. Suttkus. 1977. *Ambloplites constellatus,* a new species of rock bass from the Ozark Upland of Arkansas and Missouri with a review of western rock bass populations. Amer. Midl. Nat. 98: 147–161.

Cross, F.B. 1970. Pleistocene and recent environments of the central Great Plains. *In*: W. Dort, Jr. & J.K. Jones, Jr. (ed.) Pleistocene and Recent Environments of the Great Plains, Department of Geology, University of Kansas, Special Publication 3.

Farris, J.S. 1972. Estimating phylogenetic trees from distance matrices. Amer. Nat. 106: 645–668.

Fitch, W.M. & E. Margoliash. 1967. Construction of phylogenetic trees. Science 155: 279–284.

Gilbert, C.R. 1980. *Etheostoma variatum* (Kirtland), the variegate darter. p. 706. *In*: D.S. Lee, C.R. Gilbert, C.H. Hocutt, R.E. Jenkins, D.E. McAllister & J.R. Stauffer, Jr. (ed.) Atlas of North American Freshwater Fishes, N.C. State Mus. Nat. Hist., Raleigh.

Hocutt, C.H. 1979. Drainage evolution and fish dispersal in the central Appalachians. Geol. Soc. Amer. Bull. Part II, 90: 197–234.

Hocutt, C.H., R.F. Denoncourt & J.R. Stauffer, Jr. 1978. Fishes of the Greenbrier River, West Virginia, with drainage history of the central Appalachians. J. Biogeog. 5: 59–80.

Hocutt, C.H., R.F. Denoncourt & J.R. Stauffer, Jr. 1979. Fishes of the Gauley River, West Virginia. Brimleyana 1: 47–80.

Hocutt, C.H., R.E. Jenkins & J.R. Stauffer, Jr. 1980a. *Etheostoma osburni* (Hubbs and Trautman), the finescaled saddled darter. p. 678. *In*: D.S. Lee, C.R. Gilbert, C.H. Hocutt, R.E. Jenkins, D.E. McAllister & J.R. Stauffer, Jr. (ed.) Atlas of North American Freshwater Fishes, N.C. State Mus. Nat. Hist., Raleigh.

Hocutt, C.H., R.E. Jenkins & J.R. Stauffer, Jr. 1980b. *Etheostoma kanawhae* (Raney), the Kanawha darter. p. 659. *In*: D.S. Lee, C.R. Gilbert, C.H. Hocutt, R.E. Jenkins, D.E. McAllister, J.R. Stauffer, Jr. (ed.), Atlas of North American Freshwater Fishes, N.C. State Mus. Nat. Hist., Raleigh.

Hubbs, C.L. & V.D. Black. 1940. Percid fishes related to *Poecilichthys variatus,* with descriptions of three new forms. Occ. Pap. Mus. Zool. Univ. Mich. 416: 1–30.

Hubbs, C.L. & M.B. Trautman. 1932. *Poecilichthys osburni,* a new darter from the upper Kanawha River system in Virginia and West Virginia. Ohio J. Sci. 32: 31–38.

Jenkins, R.E., E.A. Lachner & F.J. Schwartz. 1972. Fishes of the central Appalachian drainages, their distribution and dispersal. pp. 43–117. *In*: P.C. Holt, R.A. Paterson & J.P. Hubbard (ed.) The Distributional History of the Biota of the

Southern Appalachians, Part III: Vertebrates, Virginia Polytechnic Institute, Resources Division Monograph 4.

Lachner, E.A. & R.E. Jenkins. 1971. Systematics, distribution and evolution of the chub genus *Nocomis* Girard (Pisces, Cyprinidae) of eastern United States, with descriptions of new species. Smith. Contrib. Zool. 85: 1–97.

May, B. 1975. Electrophoretic variation in the genus *Oncorhynchus*: The methodology, genetic basis, and practical applications of research and management. M. Sc. Thesis. Washington State University, Pullman. 96 pp.

Miller, R.V. 1968. A systematic study of the greenside darter *Etheostoma blennioides* Rafinesque (Pisces: Percidae). Copeia 1968: 1–40.

Nei, M. 1972. Genetic distance between populations. Amer. Nat. 106: 283–292.

Nei, M. 1973. Analysis of gene diversity in subdivided populations. Proc. Natl. Acad. Sci. U.S. 70: 3320–3323.

Page, L.M. 1981. The genera and subgenera of darters (Percidae, Etheostomatini). Occ. Pap. Mus. Nat. Hist. Univ. Kansas 90: 1–69.

Pflieger, W.L. 1971. A distributional study of Missouri fishes. Univ. Kans. Publ. Mus. Nat. Hist. 20: 225–570.

Prager, E.M. & A.C. Wilson. 1976. Congruency of phylogenies derived from different proteins. A molecular analysis of the phylogenetic position of cracid birds. J. Mol. Evol. 9: 45–57.

Raney, E.C. 1941. *Poecilichthys kanawhae,* a new darter from the upper New River system in North Carolina and Virginia. Occ. Pap. Mus. Zool. Univ. Mich. 434: 1–16.

Ross, R.D. & B.D. Perkins. 1959. Drainage evolution and distribution of fishes of the New (Upper Kanawha) River system in Virginia, Part III, records of fishes of the New River. Virginia Agr. Exp. Sta. Tech. Bull. 145: 1–34.

Selander, R.K., M.H. Smith, S.Y. Yang, W.E. Johnson & J.B. Gentry. 1971. Biochemical polymorphism and systematics in the genus *Peromyscus* sp. I. Variation in the old field mouse (*Peromyscus polionotus*). Studies in Genetics IV. Univ. Tex. Publ. 7103: 49–90.

Sneath, P.H.A. & R.R. Sokal. 1973. Numerical Taxonomy. W.H. Freeman, San Francisco. 573 pp.

Stauffer, J.R., Jr., C.H. Hocutt & C.R. Gilbert. 1980. *Etheostoma euzonum* (Hubbs & Black), Arkansas saddled darter. p. 645. *In*: D.S. Lee, C.R. Gilbert, C.H. Hocutt, R.E. Jenkins, D.E. McAllister & J.R. Stauffer, Jr. (ed.) Atlas of North American Freshwater Fishes, N.C. State Mus. Nat. Hist., Raleigh.

Swofford, D.L. 1981. On the utility of the distance Wagner procedure. pp. 25–43. *In*: V.A. Funk & D.R. Brooks (ed.) Advances in Cladistics: Proc. First Meeting Willi Hennig Soc., N. Y. Botanical Garden, New York.

Wright, S. 1978. Evolution and genetics of populations. Vol. 4. Variability within and among natural populations. University of Chicago Press, Chicago. 580 pp.

Wright, S. 1982. Character change, speciation, and the higher taxa. Evol. 36: 427–443.

Originally published in Env. Biol. Fish. 11: 85–95

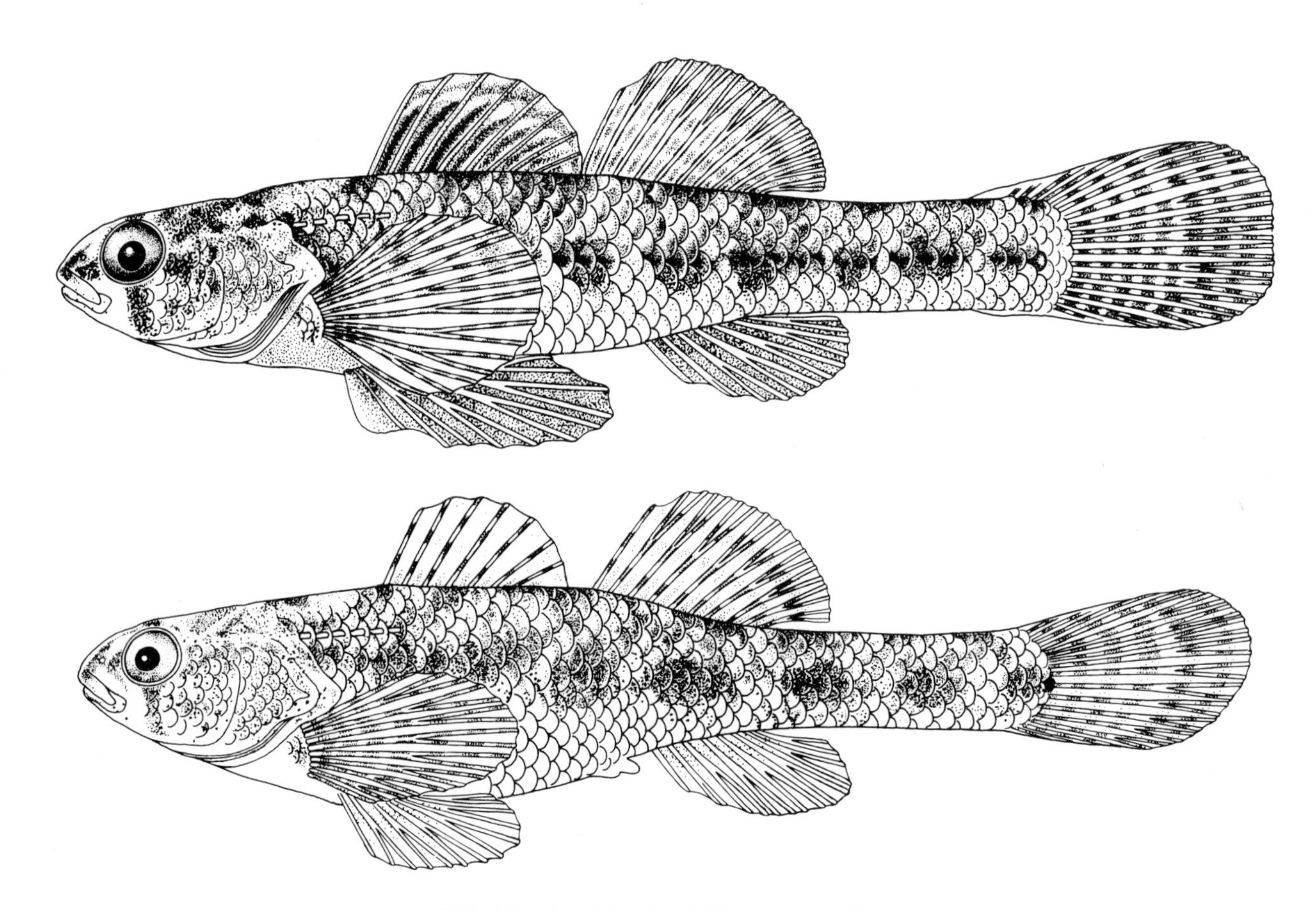

Male (above) and female of *Etheostoma proeliare*.

Ecological and evolutionary consequences of early ontogenies of darters (Etheostomatini)

Michael D. Paine
Department of Zoology, University of Guelph, Guelph, Ontario N1G 2W1, Canada

Keywords: Reproductive guilds, Yolk supply, Vitelline circulation, First feeding, Drift dispersal, Restricted gene flow, Speciation

Synopsis

The ecological classification of fishes into reproductive guilds is based on the premises that (1) reproductive styles and early ontogeny are closely related, and (2) both are correlated with the ecology of a species. A comparison of early ontogenies of logperch (*Percina caprodes*), rainbow darter (*Etheostoma caeruleum*), and fantail darter (*E. flabellare*) confirmed these premises, and provided possible explanations for diversity within the Etheostomatini. Young logperch have limited vitelline circulation, hatch while still poorly developed, and therefore must drift from oxygen rich lotic habitats to lentic habitats where small planktonic prey are available. Young rainbow and fantail darters have extensive vitelline plexuses, are well developed at transition to first feeding, and begin feeding on aquatic insects. Thus there is no necessity for a drift interval. As a result, the latter species are adapted for stream life. Interspecific differences in reproductive styles and early ontogenies may have contributed to speciation of darters by allowing partitioning of breeding sites and food resources for young. In addition, reduced drift dispersal and small stream habitation may have indirectly contributed to speciation by reducing genetic exchange among populations.

Introduction

Balon (1975a, 1981a) classified fishes into reproductive guilds, expanding on the idea of Kryzhanovsky (1949). There are two premises basic to the concept of reproductive guilds as an ecological classification: (1) Early ontogeny and reproductive styles are closely related. Adaptations of early ontogeny are correlated with factors such as spawning site and parental investment. For example, oxygen conditions of spawning grounds will be reflected in embryonic respiratory adaptations; (2) Early ontogeny is a key interval in determining the life history of a species, as even small reductions in mortality of the young can have significant effects on population numbers of animals with high mortality rates (Meats 1971). Adaptations to reduce predation or otherwise enhance survival of young fishes will therefore be strongly selected. Kryzhanovsky (1949) went further and argued that early ontogenetic adaptations '... mark the biology of adults, and define the types of migrations, invasion abilities, and limits of distribution' (p. 237). Darters (Percidae: Etheostomatini), with their diversity of reproductive styles, should show a correlated diversity of early ontogenies. Furthermore, differences in early development among species should correlate with their ecological differences as adults. I tested these predictions by comparing early ontogenies of three darters – northern logperch (*Percina caprodes semifasciata*), rainbow darter (*Etheostoma caeruleum*), and barred fantail

David G. Lindquist & Lawrence M. Page (ed.), Environmental biology of darters. ISBN 90-6193-506-7

darter (*E. flabellare flabellare*). Normal stages for all three have been previously described (Cooper 1978, 1979), but not related to ecology or evolution.

Materials and methods

This comparison came from more extensive studies of the early ontogeny of each species. Spawning logperch were captured in late May and early June 1980 in Young and Potter Creeks (Fig. 1). Fish were transported live to the laboratory where ova were stripped and fertilized in plastic dishes using the procedure of Strawn & Hubbs (1956). Spawning rainbow darters were captured in Swan Creek (Fig. 1) in May 1981, and transported live to the laboratory. Ova were stripped from females into plastic dishes, and fertilized using testes mashed in fish Ringer's solution (prepared according to Ginsburg 1963). Spawning fantail darters were captured in Lutteral Creek (Fig. 1) in June 1981, and ova stripped and fertilized in the same manner as for rainbow darters. In addition, because so few viable eggs were obtained in 1981, fantail darter nests with egg clutches were removed from Irvine Creek (Fig.

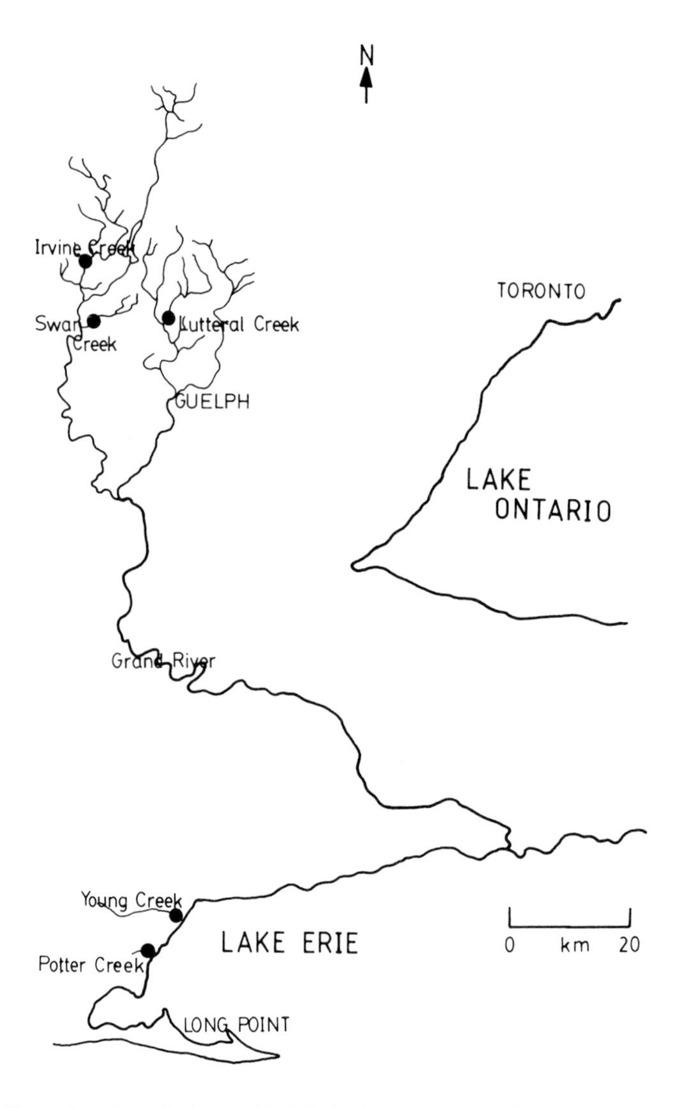

Fig. 1. Location of sites (solid circles) where spawning adults were captured.

1) in May 1982 and incubated in the laboratory.

Plastic dishes of eggs, and fantail darter egg clutches on rocks, were incubated at 19.8–20.2° C in nylon mesh breeder boxes placed in an incubation aquarium. Water was pumped from a reservoir through a biological filter into this aquarium, and returned by gravity to the reservoir. Filtered well water (10° C) (see Hodson & Sprague 1975 for chemical analysis) was added continuously (25 l h^{-1}) to the reservoir, cooling the system. Each breeder box was equipped with an airstone, and several airstones were placed in the reservoir, providing 90% or higher oxygen saturation in the breeder boxes. Airstones were removed from breeder boxes after embryos hatched, reducing turbulence, but also reducing oxygen levels to as low as 80% saturation. For all three years, a 15 light: 9 dark hours photoperiod was maintained, approximating conditions on the spawning grounds.

Because of limited egg numbers, and extensive mortality due to fungal infection, fantail darter eggs were treated twice daily for 5–10 min in a 2 ppm methylene blue bath. These baths were stopped once embryos began moving, when fungus ceased to be a problem.

Logperch and rainbow darter embryos and larvae were examined hourly on the first day but sampling frequency was reduced gradually to twice daily by 12 days. Fantail darters were sampled twice daily for the first three days, and 3–4 times weekly thereafter. Live samples of all three species were examined following McElman & Balon (1979). Neutral formalin preserved specimens were cleared and stained for bone (alizarin red) and cartilage (alcian blue) (Dingerkus & Uhler 1977).

Ages of logperch and rainbow darters are given as time from activation (see Balon 1981b). Because sampling intervals often exceeded one hour, especially for older individuals, ages were calculated from the midpoint of the sampling interval. Ages of fantail darters from 1982 were estimated by comparing them to samples from 1981 and to a few aquarium spawned eggs from 1982, and therefore should be considered approximate.

Terminology for blood vessels, bone, and cartilage follows the procedure outlined by McElman & Balon (1979). Intervals of ontogeny are as defined by Balon (1975b). For this paper, head length is the distance from the anterior tip of the snout to a line perpendicular to the body axis (notochord posterior to ventral bend) at the posteriormost margin of the auditory capsule. This measure, rather than that of Hubbs & Lagler (1958), was used because the auditory capsule, but not the opercular membrane was always visible and depicted in drawings from which measurements were made.

Results

This comparison focuses on three major areas – spawning habits and yolk provision, embryonic vitelline circulation, and skeletal development and head size at hatching and their consequences for early feeding.

The three species studied differ in reproductive styles and yolk provision. Logperch spawn in streams or lake surf zones, burying their eggs in sand or gravel (Winn 1958a, b). The adults do not guard the eggs, which are the smallest of the three species (Table 1). Rainbow darters spawn in gravel

Table 1. Cleavage egg and yolk sizes of the three species studied. Standard deviations given in parentheses. Long axis is greatest distance across the ovoid egg/yolk, short axis is greatest distance perpendicular to long axis.

Species	n	Egg size (mm)		Yolk size (mm)	
		Long axis	Short axis	Long axis	Short axis
Percina caprodes	9	1.23 (0.06)	1.19 (0.05)	0.91 (0.06)	0.88 (0.07)
Etheostoma caeruleum	16	2.18 (0.06)	1.91 (0.05)	1.66 (0.06)	1.56 (0.05)
E. flabellare	5	2.71 (0.09)	2.36 (0.19)	2.06 (0.04)	1.71 (0.05)

and stream riffles (Winn 1958a, b). The eggs are larger than those of logperch (Table 1), but are not guarded. Fantail darters spawn in stream rubble raceways or slow riffles, and males guard eggs deposited on the undersides of rocks (Lake 1936, Winn 1958a, b). The eggs are the largest of the three species (Table 1).

Vitelline circulation in the logperch embryo consisted of an unbranched subintestinal vitelline vein (Fig. 2a). In contrast, rainbow and fantail darters had vitelline plexuses formed by branching of the subintestinal vitelline vein (Fig. 2b, c). There were more branches across the yolk of the fantail darter.

Logperch hatched in 5–6 days at 20° C. Skeletal development of the head was limited, jaws were poorly developed with no teeth present (Fig. 3a). Mouth opening was not observed until 6 days 20 h after activation. Head length of Young Creek free embryos ranged from 13–15% of total length (5.9–6.2 mm); head length of Potter Creek free embryos ranged from 14–16% of total length (5.5–6.0 mm).

Logperch free embryos were pelagic, but did not show strong positive or negative phototropism. They were often observed suspended at the water surface if turbulence was reduced. Feeding on brine shrimp (*Artemia salina*) nauplii was first observed 8 days 5 h after activation. Larvae became increasingly benthic after this. Head length of Young Creek larvae was ≥15% of total length (5.7–6.2 mm); head length of Potter Creek larvae was ≥16% of total length (5.3–6.5 mm). Many logperch did not feed, and all died by age 13 days. Despite this, logperch free embryos and larvae did have smaller heads than rainbow or fantail darter free embryos or larvae.

Rainbow darters hatched at age 8–9 days, and had more head cartilage (Fig. 3b) than did logperch. Jaws were better developed, and teeth were present in most specimens at or soon after hatching. Head length of free embryos was 16–18% of total length (6.8–7.1 mm). Rainbow darters were largely benthic after hatching, and surface suspension was never observed.

A few rainbow darters had ciliates in their gastrointestinal tracts prior to age 12 days, but all specimens age 12 days or older had eaten numerous brine shrimp nauplii. Therefore age 12 days was considered the start of the larval period. Both relative head length and total length increased rapidly after feeding began. Relative head length increased from 20.7% at age 12 days 5 h to 26.4% at age 14 days 5 h; total length increased from 8.0 to 8.5 mm over the same interval.

Fantail darters from both Lutteral and Irvine Creeks hatched 12–14 days after activation. Head cartilage was extensive, jaws with teeth were well developed, rays supported by distal and proximal pterygiophores were present in medial fins, and both neural and haemal arches had formed along the notochord (Fig. 3c). Head length of Irvine Creek free embryos was 20–21% of total length (8.9–9.0 mm). Fantail free embryos were about 2 mm longer than rainbow darter free embryos, and about 3 mm longer than logperch free embryos. Fantail darter free embryos were benthic, rarely venturing into the water column.

Fantail darters began feeding 2–3 days after hatching, by which time the medial fins were differentiated. The larval period, if present at all, was therefore limited to a vestige of the pterolarval phase – the interval of advanced differentiation in unpaired fins (Balon 1975b, 1984). Total length of alevins was >9.2 mm; relative head length >20%.

Discussion

Based on the results of this study, and Winn (1958a, b), logperch are nonguarding, open substrate, rock and gravel spawners with pelagic larvae (guild A.1.2 of Balon 1981a). Logperch may spawn on sand, as well as gravel, and bury their eggs to some extent (Winn 1958a, b), but have more of the key features of guild A.1.2 (pelagic free embryos, no photophobia, limited embryonic respiratory structures) than of any other guild. Rainbow darters are nonguarding, open substrate, rock and gravel spawners with benthic larvae (guild A.1.3). The female buries into the gravel before the male mounts her (Winn 1958a, b), so the eggs may be buried deeper than logperch eggs. The difference is probably slight, and eggs are not buried to the extent that eggs of typical broodhiders such as trout or charr are. Fantail darters are guarding hole nesters (guild B.2.7).

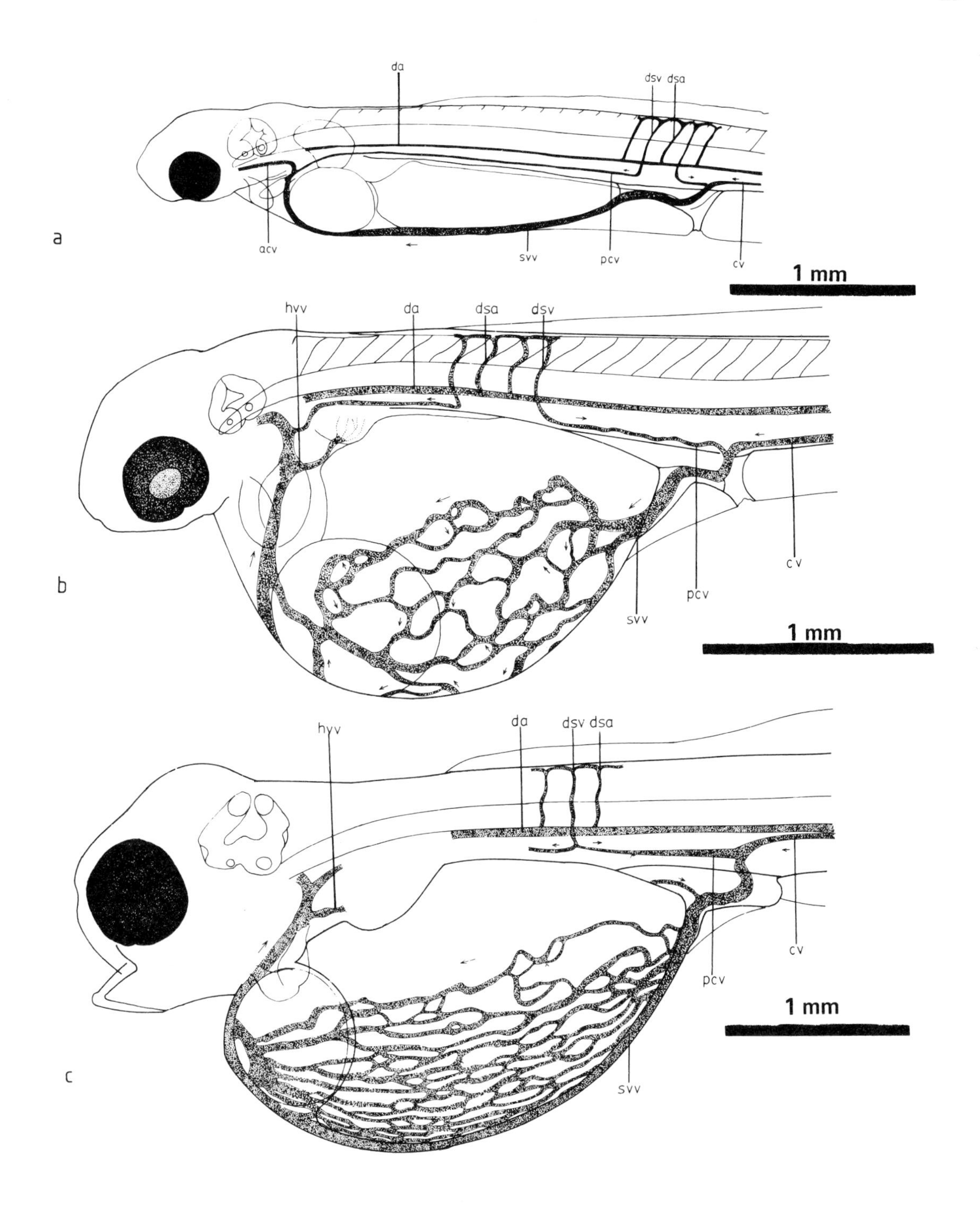

Fig. 2. Left lateral views of vitelline circulation in the three species studied. a – Young Creek logperch (*Percina caprodes*), age 6 d 8 h 50 min, b – rainbow darter (*Etheostoma caeruleum*), age 5 d 13 h 50 min, after excision from the egg envelope, c – Irvine Creek fantail darter (*E. flabellare*), age 11 d, after excision from the egg envelope (acv – anterior cardinal vein, cv – caudal vein, da – dorsal aorta, dsa – dorsal segmental artery, dsv – dorsal segmental vein, hvv – hepatic vitelline vein, pcv – posterior cardinal vein, svv – subintestinal vitelline vein).

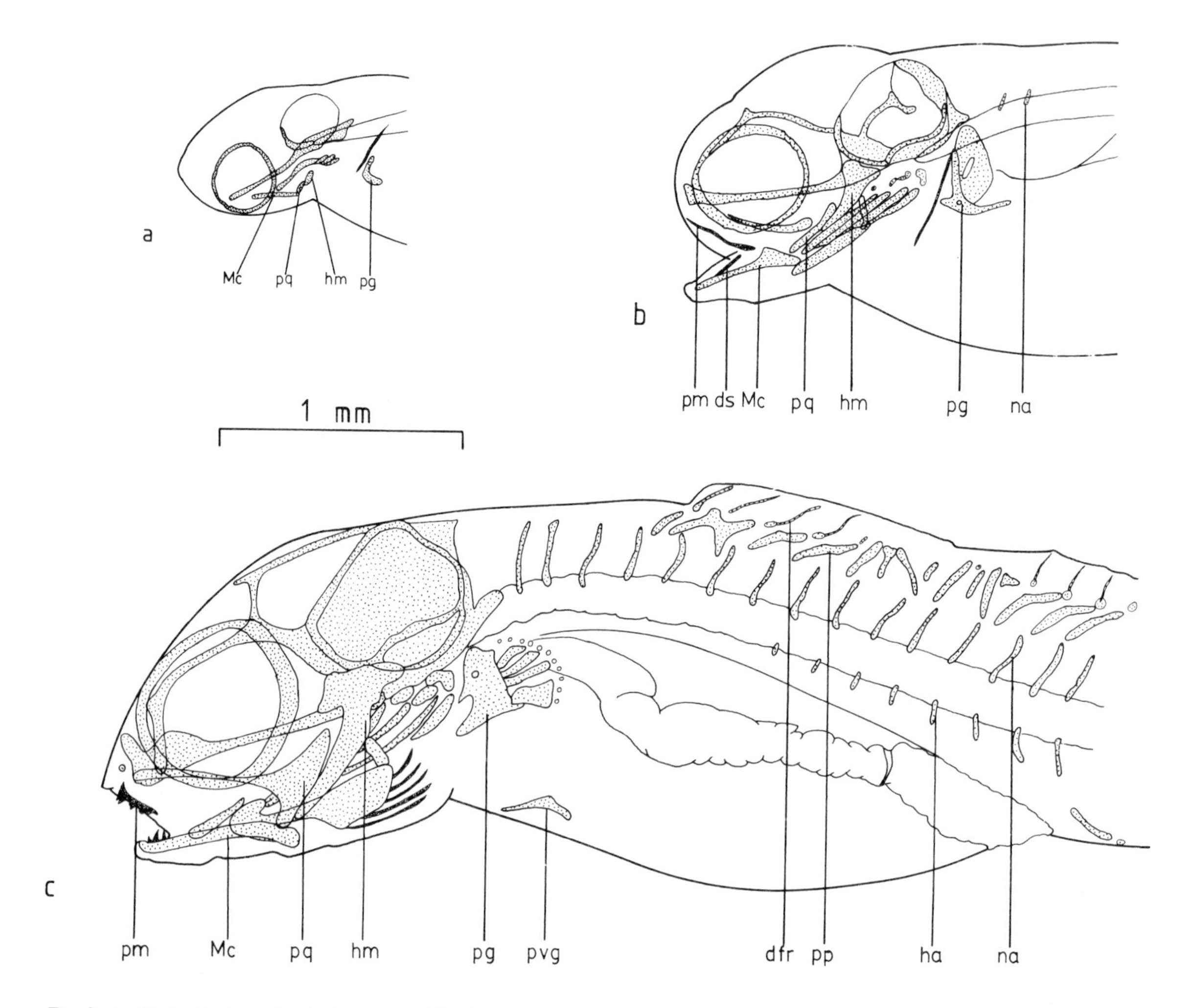

Fig. 3. Left lateral views of anterior skeletal development of free embryos of the three species studied. Cartilage – light stipple, bone – heavy stipple. a – Young Creek logperch (*Percina caprodes*), age 5 d 23 h 50 min, b – rainbow darter (*Etheostoma caeruleum*), age 9 d 7 h 20 min, c – Lutteral Creek fantail darter (*E. flabellare*), age 13 d 17 h. Note the fusion of some first dorsal proximal pterygiophores in the fantail darter, an abberant condition (dfr – dorsal fin ray, ds – dentosplenial, ha – haemal arch, hm – hyomandibular, Mc – Meckel's cartilage, na – neural arch, pg – pectoral girdle, pm – premaxillary, pp – proximal pterygiophore, pq – palatoquadrate, pvg – pelvic girdle).

How are the ontogenies of these three species related to their reproductive styles? Kryzhanovsky (1949) considered oxygen conditions at the spawning site to be correlated with embryonic respiratory adaptations such as vitelline plexuses. An extensive vitelline plexus exposes a large area of blood at the yolk surface, enhancing oxygen diffusion. For the three species studied, plexus size is correlated with embryo size, spawning temperature, and spawning site.

There was an increase in embryo size (primarily depth and width) from logperch to rainbow darter to fantail darter and a corresponding increase in extent of the vitelline plexus. Larger embryos have reduced surface area:volume, and therefore reduced oxygen diffusion across the body. Increased vitelline plexus size also appears correlated with increased spawning and natural incubation temperatures. Logperch spawning began in Young and Potter Creeks at 12–14° C, rainbow darter spawning began in Swan Creek at 16–17° C, and fantail darter spawning began in Irvine Creek at 17–20° C. Winn

(1958a) observed that fantail darters began spawning a month later, and therefore presumably at higher temperatures, than logperch or rainbow darters. Oxygen concentrations are of course lower at higher temperatures. The hypothesized relationship between natural incubation temperature and extent of vitelline plexus is only valid if the latter is a genetically determined, and not environmentally induced characteristic.

Logperch embryos develop on or near the substrate in lotic conditions, where oxygen levels are high, and therefore do not require extensive respiratory networks. Walleye (*Stizostedion vitreum*) embryos develop under similar conditions, also have an unbranched subintestinal vitelline vein (McElman & Balon 1979), despite being larger than logperch embryos. In contrast, fantail darter embryos, and the much smaller embryos of the johnny darter (*Etheostoma nigrum,* personal observation), develop in closely packed rock nests in slower waters, where oxygen levels are probably lower. Both species have extensive vitelline plexuses. Thus, there appears to be a correlation, independent of embryo size, between extent of vitelline plexus and oxygen levels at the spawning site. Rainbow darters, however, do not fit this pattern, as embryos develop under the same conditions as logperch embryos, but have larger vitelline plexuses.

Another characteristic of reproductive styles with ontogenetic and ecological effects is extent of yolk provision. Embryos with large yolks can remain within egg envelopes longer, and hatch as larger, better formed, and less vulnerable (to predation) fish. Interspecific differences in absolute size, head size, and skeletal development at hatch and first feeding were evident in this study, and are correlated with size of prey at first feeding.

Logperch larvae are probably unable to begin feeding on aquatic insects, because of their small poorly developed heads. They feed on microcrustaceans until they reach 25 mm standard length (Turner 1921). Fantail darters have a reduced or absent larval period, and alevins, with large well developed heads, feed on aquatic insects, not microcrustaceans (Turner 1921, Paine 1979). Rainbow darter larvae are probably capable of feeding on aquatic insects as well, and microcrustaceans are of limited importance in the diet of fish 15–20 mm standard length (Turner 1921). Unfortunately the diet of smaller specimens has not been reported.

How do logperch find small planktonic food, rare in the lotic conditions of the spawning sites? Free embryos were pelagic, although distinct positive phototaxis was not observed. Pelagic free embryos would be carried downstream until they reached lentic conditions (lakes, pools, stream edges, eddies, bays, backwaters). There, plankton (algae, rotifers, crustaceans) is most common (Hynes 1970), and feeding may begin. Drift may also conserve energy, as free embryos are poorly developed skeletally and may therefore be unable to hold position or swim upstream where current exists. Logperch have a 'moderately long free swimming' interval during development (Hubbs & Strawn 1957, p. 47), and larvae have been captured in Lake Erie plankton tows (Cooper 1978), providing further evidence for the existence of a pelagic interval.

Other *Percina* species (Table 2), walleye (McElman & Balon 1979), and pike-perch (*Stizostedion lucioperca*) (Belyy 1972) are pelagic after hatching. In contrast, rainbow darter, fantail darter, and most *Etheostoma* species examined to date are benthic after hatching (Table 2). The presence or absence of a drift interval has some potential ecological consequences. If the drift interval is an adaptation for locating plankton, then species with a drift interval are not well adapted for life in small, low order streams. Plankton is generally more abundant further downstream (Hynes 1970). If most *Percina* species have a drift interval (as Table 2 suggests), then the presence of a drift interval is correlated with habitation of larger streams, rivers and lakes (Table 3). This correlation would be more plausible if a drift interval is associated with larvae incapable of feeding on aquatic insects, as it is for logperch. Note that downstream movement of young must be offset by upstream migration of juveniles or adults. Therefore, I would predict that species with drift intervals undertake more extensive spawning migrations than those without.

A reduced drift interval (Table 2) and small stream habitation (Table 3) seem to be characteris-

Table 2. Duration of drift intervals for Etheostomatini species. References in parentheses.

Long drift interval	Reduced or absent drift interval
Percina – *P. caprodes* (1,4)	*Percina* – none reported
P. macrocephala (9)	*Etheostoma* – *E. caeruleum* (1)
P. maculata (6)	*E. flabellare* (1)
P. sciera (4)	*E. fonticola* (4)
P. tanasi (8)	*E. lepidum* (4)
	E. nigrum (2, 3)
Etheostoma – *E. exile* (3)	*E. perlongum* (7)
	E. spectabile (4)
	E. squamiceps (5)

References: 1 – this study, 2 – personal observation, 3 – Balesic 1971, 4 – Hubbs & Strawn 1957, 5 – Page 1974, 6 – Petravicz 1938, 7 – P. Shute in Lee et al. 1980, 8 – Starnes 1977, 9 – W.C. Starnes (personal communication).

tic of most *Etheostoma* species. A reduced drift interval will only be an adaptation for small stream habitation if drift to plankton food sources is not required. This will be the case for species (e.g. fantail darter) capable of using aquatic insects as first prey. Species with a reduced drift interval and large young, while adapted for small streams, are not therefore any less adapted for larger streams, rivers, or lakes. Aquatic insects are universally available, and fantail darters, for example, are found in Lake Erie (Scott & Crossman 1973).

Interspecific differences in reproductive styles and early ontogenies may have directly and indirectly contributed to the present day diversity of darters. Direct effects would include mechanisms for partitioning resources or utilizing previously unused resources, indirect effects would include mechanisms increasing genetic isolation among stocks.

Table 3. Distribution of *Percina* and *Etheostoma* species according to waterway size (after Lee et al. 1980).

Genus	Number of species			
	Small to medium streams[1]	Larger waterways[2]	Both[3]	Unclassified[4]
Percina	7	13	2	8
Etheostoma	38	20	10	22

[1] Includes 'springs', 'spring runs', 'headwater streams', 'creeks'.
[2] Includes 'medium to large streams', 'rivers', 'main channels', 'lakes'.
[3] Species found in both small to medium streams and larger waterways.
[4] Stream or river size preferred not indicated.

The development of embryonic adaptations, such as extensive vitelline plexuses, may have allowed some darter species to exploit breeding sites other than the riffle gravel beds used by most primitive (*Percina*) species. Constanz (1979) and Lindquist et al. (these proceedings) provide evidence that breeding sites are a limited resource for some darter species. Interspecific differences in spawning time may allow partitioning of breeding sites, and temporarily segregate the young of species consuming the same food resource. A reduced drift interval and large larvae or alevins may have permitted some species to colonize smaller streams. Certainly the differences in spawning time, breeding site, and size at first exogenous feeding between rainbow and fantail darters offer as good or better an explanation for the coexistence of these two species than the adult morphological and behavioural differences suggested by Paine et al. (1982).

Reduced drift dispersal may have indirectly led to speciation by reducing genetic exchange among stocks. Reduced dispersal of young has been correlated with speciation in marine invertebrates (e.g. Scheltema 1971, 1978, Hansen 1978, Strathmann 1978) and marine fishes (e.g. Rosenblatt 1963, Barlow 1981). Grassle & Grassle (1978) showed that the percentage of loci that were polymorphic was greater for marine polychaetes with shorter larval dispersal intervals, and this experiment could easily be repeated comparing darter species with and without drift dispersal. Small stream habitation, which I have correlated with reduced drift dispersal, will further enhance genetic isolation (Mahon 1981). Two stocks in the headwaters of two different rivers in the same drainage basin are obviously separated by a greater distance than stocks downstream from either. The fact that there are more species of *Etheostoma* than of *Percina* could be partially the result of reduced drift dispersal and stream habitation.

In conclusion, the reproductive styles and early ontogenies of the three species studied appear correlated with their ecology. Furthermore, interspecific differences in reproductive styles and early ontogenies may have had direct and indirect effects on the present day diversity of darters. If the premises of the concept of reproductive guilds are valid, then studies of the early ontogenies of darters would contribute substantially to our understanding of darter ecology. Vrba's (1980) effect hypothesis provides a conceptual basis for linking reproductive specialization and speciation, and may be applicable to evolutionary studies of darters.

Acknowledgments

Eugene Balon provided guidance, critical comments, and funding from a Natural Sciences and Engineering Research Council (NSERC) grant. Robin Mahon contributed greatly to many of the ideas put forward in this paper. Jim Baker, David Noakes, and Martyn Obbard read and criticized the manuscript. Their efforts are appreciated, and any problems that remain are my own responsibility. I would like to thank Jim Bowlby, Joan Cunningham, Roy Danzmann, Moira Ferguson, Colin Macdonald, Willem Pot, Rick Procter, Kevin Reid, Dwight Watson, and Marilyn White for their help in the field, and George Dixon and Jim McElman for their advice on constructing incubation systems. Cam Portt arranged for permission to sample local streams from the Cambridge office of the Ontario Ministry of Natural Resources (OMNR); David Read of the Simcoe OMNR was especially helpful in arranging for collection of logperch. Robert Black and Robert Woodward kindly allowed me to sample on their properties. I would especially like to thank Larry Halyk for all the hours he has spent on all aspects of this study, and NSERC for providing support through a Postgraduate Scholarship.

References cited

Balesic, H. 1971. Comparative ecology of four species of darters (Etheostomatinae) in Lake Dauphin and its tributary, the Valley River. M.Sc. Thesis, Univ. Manitoba, Winnipeg. 77 pp.

Balon, E.K. 1975a. Reproductive guilds of fishes: a proposal and definition. J. Fish. Res. Board Can. 32: 821–864.

Balon, E.K. 1975b. Terminology of intervals in fish development. J. Fish. Res. Board Can. 32: 1663–1670.

Balon, E.K. 1981a. Additions and amendments to the classification of reproductive styles in fishes. Env. Biol. Fish. 6: 377–389.

Balon, E.K. 1981b. Saltatory processes and altricial to precocial forms in the ontogeny of fishes. Amer. Zool. 21: 573–596.

Balon, E.K. 1984. Patterns in the evolution of reproductive styles in fishes. pp. 35–53. *In*: C.W. Potts & R.J. Wootton (ed.) Fish Reproduction: Strategies and Tactics, Academic Press, London.

Barlow, G.W. 1981. Patterns of parental investment, dispersal and size among coral-reef fishes. Env. Biol. Fish. 6: 65–85.

Belyy, N.D. 1972. Downstream migration of the pike-perch *Lucioperca lucioperca* (L.) and its food in the early development stages in the lower reaches of the Dnieper. J. Ichthyol. 12: 465–472.

Constanz, G.D. 1979. Social dynamics and parental care in the tessellated darter (Pisces: Percidae). Proc. Acad. Nat. Sci. Phila. 131: 131–138.

Cooper, J.E. 1978. Eggs and larvae of the logperch, *Percina caprodes* (Rafinesque). Amer. Midl. Nat. 99: 257–269.

Cooper, J.E. 1979. Description of eggs and larvae of fantail (*Etheostoma flabellare*) and rainbow (*E. caeruleum*) darters from Lake Erie tributaries. Trans. Amer. Fish. Soc. 108: 46–56.

Dingerkus, G. & L.D. Uhler. 1977. Enzyme clearing of alcian blue stained whole small vertebrates for demonstration of cartilage. Stain Technol. 52: 229–232.

Ginsburg, A.S. 1963. Sperm-egg association and its relationship to the activation of the egg in salmonid fishes. J. Embryol. Exp. Morphol. 11: 13–33.

Grassle, J.F. & J.P. Grassle. 1978. Life histories and genetic variation in marine invertebrates. pp. 347–364. *In*: B. Battaglia & J.A. Beardmore (ed.) Marine Organisms, Plenum Press, New York.

Hansen, T.A. 1978. Larval dispersal and species longevity in Lower Tertiary gastropods. Science 199: 885–887.

Hodson, P.V. & J.B. Sprague. 1975. Temperature-induced changes in acute toxicity of zinc to Atlantic salmon (*Salmo salar*). J. Fish. Res. Board Can. 32: 1–10.

Hubbs, C. & K. Strawn. 1957. Survival of F_1 hybrids between fishes of the subfamily Etheostomatinae. J. Exp. Zool. 134: 33–62.

Hubbs, C.L. & K.F. Lagler. 1958. Fishes of the Great Lakes Region. Rev. ed. Cranbrook Inst. Sci. Bull. No. 26. 213 pp.

Hynes, H.B.N. 1970. The ecology of running waters. Univ. Toronto Press, Toronto. 555 pp.

Kryzhanovsky, S.G. 1949. Eco-morphological principles of develoment among carps, loaches and catfishes. Tr. Inst. Morph. Zhiv. Severtsova 1: 5–332. (In Russian). (Part 2, Ecological groups of fishes and patterns of their distribution, pp. 237–331, Transl. from Russian by Fish. Res. Board Can. Transl. Ser. No. 2945, 1974).

Lake, C.T. 1936. The life history of the fan-tailed darter *Catonotus flabellaris flabellaris* (Rafinesque). Amer. Midl. Nat. 17: 816–830.

Lee, D.S., C.R. Gilbert, C.H. Hocutt, R.E. Jenkins, D.E. McAllister & J.R. Stauffer, Jr. 1980. Atlas of North American freshwater fishes. N.C. State Mus. Nat. Hist., Raleigh. 867 pp.

Mahon, R. 1981. Patterns in the fish taxocenes of small streams in Poland and Ontario. Ph.D. Thesis, University of Guelph, Guelph. 297 pp.

McElman, J.R. & E.K. Balon. 1979. Early ontogeny of walleye *Stizostedion vitreum,* with steps of saltatory development. Env. Biol. Fish. 4: 309–348.

Meats, A. 1971. The relative importance to population increase of fluctuations in mortality, fecundity and the time variables of the reproductive schedule. Oecologia 6: 223–237.

Page, L.M. 1974. The life history of the spottail darter, *Etheostoma squamiceps,* in Big Creek, Illinois, and Ferguson Creek, Kentucky. III. Nat. Hist. Surv. Biol. Notes 89: 1–20.

Paine, M.D. 1979. Feeding relationships of four species of darters (*Etheostoma*) in Irvine Creek, Ontario. M.Sc. Thesis, University of Waterloo, Waterloo. 163 pp.

Paine, M.D., J.J. Dodson & G. Power. 1982. Habitat and food resource partitioning among four species of darters (Percidae: *Etheostoma*) in a southern Ontario stream. Can. J. Zool. 60: 1635–1641.

Petravicz, W.P. 1938. The breeding habits of the black-sided darter, *Hadropterus maculatus* Girard. Copeia 1938: 40–44.

Rosenblatt, R.H. 1963. Some aspects of speciation in marine shore fishes. Syst. Assoc. Publ. 5: 171–180.

Scheltema, R.S. 1971. Larval dispersal as a means of genetic exchange between geographically separated populations of shallow-water benthic marine gastropods. Biol. Bull. 140: 284–322.

Scheltema, R.S. 1978. On the relationship between dispersal of pelagic veliger larvae and the evolution of marine prosobranch gastropods. pp. 303–322. *In*: B. Battaglia & J.A. Beardmore (ed.) Marine Organisms, Plenum Press, New York.

Scott, W.B. & E.J. Crossman. 1973. Freshwater fishes of Canada. Fish. Res. Board Can. Bull. 183. 966 pp.

Starnes, W.C. 1977. Ecology and life history of the endangered snail darter, *Percina tanasi* Etnier. Tenn. Wildlife Resources Agency Tech. Rept.: 77–52.

Strathmann, R.R. 1978. The evolution and loss of feeding larval stages of marine invertebrates. Evolution 32: 894–906.

Strawn, K. & C. Hubbs. 1956. Observations on stripping small fishes for experimental purposes. Copeia 1956: 114–116.

Turner, C.L. 1921. Food of the common Ohio darters. Ohio J. Sci. 22: 41–62.

Vrba, E.S. 1980. Evolution, species and fossils: how does life evolve? S. Afr. J. Sci. 76: 61–84.

Winn, H.E. 1958a. Observations on the reproductive habits of darters (Pisces-Percidae). Amer. Midl. Nat. 59: 190–212.

Winn, H.E. 1958b. Comparative reproductive behavior and ecology of fourteen species of darters (Pisces-Percidae). Ecol. Monog. 28: 155–191.

Originally published in Env. Biol. Fish. 11: 97–106

Selection of sites for egg deposition and spawning dynamics in the waccamaw darter

David G. Lindquist, John R. Shute[1], Peggy W. Shute[1] & L. Michael Jones
Department of Biological Sciences, University of North Carolina-Wilmington, NC 28406, U.S.A.

Keywords: Experimental spawning cover, Breeding season, Nest choice, Nest fidelity, Nest egg number, Nest quality, Male and female size, Fish, Percids

Synopsis

We provided 93 experimental spawning covers for the waccamaw darter. We grouped the covers (3 sizes of slate and one of concave tile) in three arrangements at six Lake Waccamaw locations to separate the variables of water depth, distance from shore, cover density and cover type. Tag returns of marked males suggest low fidelity for nest sites. Egg production under the 3 different sizes of slate was not significantly different. Egg production under the tile was significantly less than that under the slates. Egg production was significantly higher off the undeveloped southeastern shore in 2 m of water and lowest at the shallowest location with the highest experimental cover density. The number of eggs in nest is positively correlated with male size. We conclude that medium size slate covers placed in a linear arrangement in 2 m of water on a mixed sand bottom result in the highest egg production for the waccamaw darter.

Introduction

The sites of egg deposition in darters have been summarized for approximately one-third of the species (Page 1983, Page et al. 1982). Darters bury their eggs in the substrate, attach them to the tops or sides of objects (plants or rocks) or cluster or clump them underneath an object (rock or log). In the former two modes, eggs are abandoned after deposition; in the latter two, they are guarded by the male. The factors that govern nest site selection are unknown for most species. Winn (1956) experimented in the laboratory with egg site choice for *Etheostoma blennioides, E. caeruleum* and *E. spectabile,* species that abandon their eggs after spawning. No nest site selection experiments have been performed on egg guarding species (but see Speare 1965 and Constanz 1979 for related observations on *E. nigrum* and *E. olmstedi).*

The waccamaw darter, *E. perlongum,* is an egg-clustering annual species (see Lindquist et al. 1981 and Shute et al. 1982 for breeding behavior and early life history) endemic to Lake Waccamaw, North Carolina (Hubbs & Raney 1946, Bailey 1977). In connection with a survey to assess the status and conservation of the waccamaw darter, we placed artificial nesting covers in Lake Waccamaw. Objectives were to: (1) determine the size, shape and lake placement of artificial spawning covers for maximizing nest egg numbers; (2) investigate spawning dynamics and egg site selection; and (3) assess the effect of madtom catfishes, which share the artificial nesting covers, on darter egg numbers in nest.

[1] Present address: Department of Zoology, University of Tennessee, Knoxville, TN 37916, U.S.A.

David G. Lindquist & Lawrence M. Page (ed.), Environmental biology of darters. ISBN 90-6193-506-7

Methods and materials

During the first week in March 1981 a total of 93 spawning covers were placed in six different areas of Lake Waccamaw lacking logs, sticks, rocks or other potential spawning sites (Table 1). Two stations (2 and 4) were located off the northeast shore, three (1, 3 and 6) off the southeast shore and one (5) off the south shore. Three of the covers (types 1, 2 and 3; Table 1) were flat slate, 1.5 to 2 cm thick and 21.6, 29.2 and 44.5 cm square, respectively. Cover type 4 (Table 1) was a 36 x 23 cm concave terracotta roofing tile. The slates were elevated 2 to 3 cm on one side by a peg driven into the sand. The tiles were placed concave side down with one end buried leaving a single entrance at the opposite end. Covers in linear arrangements (Table 1) were 1 m apart with elevated ends facing a common direction. One each of the type 3 covers was used at stations 1, 2 and 3 and were followed by alternating types 1, 2 and 4 (5 each). Station 4 was identical, except for the absence of the type 3 cover. Station 5 was a rectangular cluster (4 slates wide and 5 slates long) of 20 alternating type 1 and 2 covers. The covers were separated by 2 to 20 cm and also were elevated 2 to 3 cm on one side. The side elevated was chosen by random number designation.

We sampled each spawning cover five to six times at regular intervals for darters and darter eggs until spawning had terminated on June 11, 1981. Other fishes (mainly the tadpole madtom, *Noturus gyrinus,* and the undescribed broadtail madtom, *Noturus* sp.) under the slates were also enumerated.

Darters were captured under cover types 1, 2 and 4 with a capture box (Downhower & Brown 1977) made of pine (2 x 2 x 30 cm) forming an open frame cube (34 cm sides). Otherwise, handnets were used to capture darters. Three sides of the capture box were covered with net (2 x 5 mm mesh) and the other side had a 85 cm long bag of 1 mm mesh net attached. The top and bottom were left open. The open bottom was weighted on two sides and three 18 cm aluminium nails could be pushed through three sides of the bottom frame into the bottom, thus anchoring the capture box to the sediment. Once the capture box was placed over the spawning cover, (without disturbing neighboring covers) the cover was lifted through the open top of the capture box exposing darters (and other fishes) that could then be forced into the net bag. Covers with egg clusters adhering to their undersides were photographed next to a ruler and replaced to their original positions. Darters were sexed and measured; marked (during initial half of spawning season only) by injecting acrylic dye (Thresher & Gronnel 1978) under a few scales (cheek, nape, under dorsal or anal fins and caudal peduncle), and released under their original spawning covers where they resumed normal behavior.

The photo transparencies of the egg clusters were used to determine egg numbers by direct count (small clusters) or by estimation (large clusters). Large egg numbers were estimated by projecting the transparency onto a small screen and tracing the outline of the cluster while taking 2 to 4 egg density samples. The area of the egg cluster was measured with a planimeter and then multiplied by the mean density estimate to arrive at the total number of eggs. A direct count check of our estimation technique gave a mean accuracy of 86% (N = 30). Data analysis was assisted with the

Table 1. Protocol for experimental spawning cover arrangements. See text for description of cover types.

Station	Water depth (m)	Offshore (m)	Cover type(s)	Cover arrangement	Total covers
1	0.5	300	1, 2, 3, 4	Linear	16
2	1	200	1, 2, 3, 4	Linear	16
3	2	1000	1, 2, 3, 4	Linear	16
4	2	600	1, 2, 4	Linear	15
5	0.2	10	1, 2	Cluster	20
6	0.5	350	4	Linear	10

Statistical Analysis System (SAS) computer programs.

Results

No marked females (N = 35) and only 24 (15%) of the marked males were recaptured under the spawning covers. No darter was recaptured more than once. Marked and recaptured males were 'at large' (between captures) an average of 26 days (range 9–66 days) or 1.5 sampling intervals (range 1–4 intervals) and had moved an average of 2.6 m. Only three of the recaptured males were recaptured under their original spawning cover.

Egg production for the four spawning covers located at stations 1, 2, 3 and 4 was greater on the three flat slates (cover types 1, 2 and 3) than on concave tile (Fig. 1). Single peaks of egg production for the four covers occur during each of the months of March, April and May with April having the greatest egg production except for cover type 3 which has a peak of nearly 5000 in May (Fig. 1). Duncan's multiple range test (DMRT) for mean egg production by cover type across the spawning season gives means with non-significant differences of 1059, 1560 and 1754 eggs for types 1, 2, and 3, respectively. The concave tile, type 4, has a mean of 474 eggs which is significantly less ($p<0.05$) than the three flat slates.

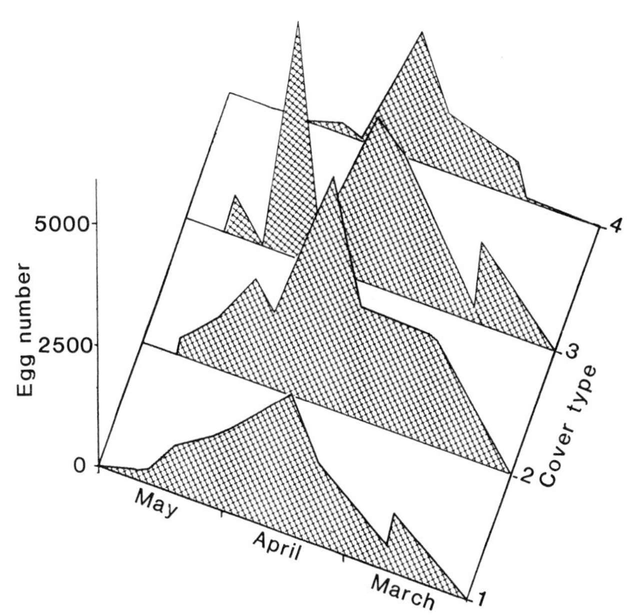

Fig. 1. Egg production (mean egg numbers) by observation date on the four spawning cover types (including covers with no eggs) combined for stations 1, 2, 3 and 4. Cover types 1, 2 and 3 are 23, 29 and 44 cm slates, respectively, and type 4 is the concave tile.

Egg production for station 6, which has concave tiles only, compares favorably with egg production for concave tiles at station 3 where darters could choose between the four cover types (DMRT, Table 2). However, egg production for concave tiles at stations 1, 2 and 4 was significantly less than that at station 6 (DMRT, Table 2). These results suggest that egg production under concave tiles can be increased by providing them as the only spawning cover (e.g., station 6, Table 2).

A comparison of mean egg production at all six stations for all cover types shows that station 3, 2 m in depth and 1000 m off the undeveloped southeastern shore, is significantly ($p<0.05$) most productive (DMRT, Table 3). Station 6, consisting of linear tiles, is also significantly more productive than station 5, having the cluster of slates. There are no significant ($p>0.05$) differences for mean egg production for stations 1, 2, 4 and 5 (DMRT, Table 3). Station 5, the shallow (0.2 m) cluster, has the lowest mean egg production.

A linear regression of nest egg numbers on the standard length (SL) of the attendant male for all samples shows a positive correlation ($r = 0.22$, $N = 222$), that is significantly different from zero ($p = 0.001$). The regression line predicts 1000 ± 400 (= 95% Confidence Limits) eggs for a 55 mm male, 1500 ± 200 eggs for 68 mm and 2000 ± 300 eggs for a 76 mm male. Mean male standard length for each sample date (all samples) regressed over the spawning period (March-June) indicates a significant negative correlation ($r = -0.82$, $p = 0.002$ and $N = 11$). The regression line predicts a mean SL of 71 ± 2 mm for March, 68 ± 1 for April, 65 ± 2 mm for May and 62 ± 3 mm for June. A comparison of mean male SL for the four cover types at stations 1, 2, 3 and 4 (DMRT, Table 4), shows significantly larger males tending nests under the medium size slates when compared to the smaller and larger slates. Male size for tiles is not significantly different

Table 2. Egg production for the concave tile (type 4) compared for stations 1, 2, 3, 4 and 6. Means followed by the same letter are not significantly different ($p > 0.05$). Means with different letters are significantly different ($p < 0.05$).

Station	Mean eggs	N
3	1282 A	16
6	1147 A	35
2	469 B	30
1	345 B	30
4	118 B	25

Table 3. Comparison of mean egg production for the six stations. See Table 2 for explanation of letters following means.

Station	Mean eggs	N
3	2474 A	58
6	1147 B	35
2	811 B C	96
4	740 B C	73
1	688 B C	89
5	547 C	100

Table 4. Comparison of mean male standard length (SL) for the four cover types. See Table 2 for explanation of letters following means.

Cover	Area (cm^2)	Male SL	N
2	853	68.5 A	57
4	828	65.7 A B	30
1	466	65.2 B	54
3	1976	65.0 B	8

from male size for slates.

A similar analysis for females indicates no significant relationship between nesting female size (SL) and nest egg number ($r = -0.19$, $p = 0.23$ and $N = 42$) or cover type or size. A marginally significant ($p = 0.05$) positive correlation exists for mean female size (SL) by observation date regressed over the spawning period ($r = 0.66$, $N = 9$). Predicted female size for March is 48 ± 5 mm and for May is 55 ± 3 mm. A cubic regression (best fit, SAS Graphics) of mean egg numbers for each date (all stations) over the spawning period predicts peak egg production (1725 ± 500 eggs) for mid-April, the middle of the spawning season.

In order to assess the effect of madtom catfishes (*Noturus gyrinus* and *N.* sp.) on darter egg numbers, we compared mean egg numbers for cover samples without madtoms ($N = 347$) with mean egg numbers for samples with one or more madtoms ($N = 76$). The means, 1046 and 866, respectively, were not significantly ($p = 0.35$) different ($t = 0.93$, $df = 105$).

Discussion

The low (15%) recapture of marked nesting males suggests that males have a high turn-over rate at the artificial nesting areas. Moreover, since only three of the twenty-four recaptured males remained under their original spawning cover, males also may have a low fidelity for specific artifical spawning covers. One reason for the high turnover of males at spawning stations may be that as available nesting surface is occupied by eggs, males abandon their nest covers in search of new (unused) cover. As embryos hatch in five to seven days (Shute et al. 1982) and nest surface is again available, new males move in and begin nesting. Many of the artificial nest covers were completely covered with multiple female spawnings resulting in huge nest egg numbers, e.g., maximum = 8097 eggs for cover type 2. Constanz (1979) observed that the large dominant males of *E. olmstedi,* after fertilizing eggs under one rock, often moved to other spawning rocks which offered more uncovered spawning surface available to spawning females. Finally, mortality due to handling during our marking procedure cannot be eliminated as a source of the low recaptures.

It appears that the flat surfaces of the slates offer superior nest surface quality when compared to the concave surface of the tiles. Observations on other egg-clustering species supports this idea (Page & Burr 1976, Page et al. 1981). Concave surfaces are probably difficult to spawn under because darters have difficulty adpressing the urogenital opening against the curved spawning surface. Male waccamaw darters can and do easily increase the height of flat or convex spawning surfaces (allowing better urogenital adpression) by digging out the sand from under the cover (Lindquist et al. 1981). This adjustment is more difficult under the convace titles. Nevertheless, some male darters were successful in producing large egg numbers under the tiles, e.g., maximum 7354 eggs for cover type 4 at station 3.

Based on our limited number of observations for the large (44.5 cm^2) slates, we suggest that these do not offer any significant increase in nest quality and egg numbers, despite a more than twofold increase in surface area. Large nesting males and large egg clusters were more often associated with the medium size slate. This may be due to the decrease in the angle with the horizontal associated with large slates (5° for the larger slates and 10° and 15° for the medium and small slates, respectively). Speare (1965) reported that his spawning plates set at 10° were most frequently used by *E. nigrum* males. In order to be more effective, our large slates may need an additional elevation of 2 to 5 cm. Even with the darter's capability to dig out sand and adjust the cover height, this large slate area (1976 cm^2) may be too much for a small fish to adjust.

Station 3, located farthest from the shore, was the most productive. Previous study by Lindquist et al. (1981) indicated that no natural cover occurs offshore and that males undergo an offshore to onshore migration in March, thus enabling rapid colonization of the onshore stations. In addition, station 3 is off the relatively undisturbed and undeveloped southeastern shore which may contribute to better water and sediment quality.

Our results show that large males hold the artificial nest sites early in the spawning season and then are gradually replaced by smaller males. Larger males also are associated with nests having larger egg numbers. This is expected if suitable spawning cover is limited. Large males win the early aggressive encounters for nest sites and females choose these early nest sites because they offer more uncovered spawning surface (Constantz 1979). Large males may depart after available nest area is used and then search for more nest area (Constantz 1979), or they may be dying due to poor condition (Shute et al. 1982).

The maximum number of eggs in a single nest for *E. perlongum* reported by us here is considerably larger than that for other reported darter nests. We believe this is a direct result of our concentration of nest sites that effects a lek-like system that maximizes nest egg production. Conservation practices for *E. perlongum* should include medium size slate covers placed in a linear arrangement in 2 m of water on a mixed sand bottom. The choice of medium size slates is most effective in egg production.

Acknowledgements

This study was made possible by grant-in-aid funds under Section 6 of the Endangered Species Act of 1973 (PL 93–205) administered by the North Carolina Wildlife Resources Commission. We thank R. Hylton, D. Goley and G. Ogburn for assistance.

References cited

Bailey, J.R. 1977. Freshwater fishes. pp. 265–98. *In:* J.E. Cooper, S.S. Robinson & J.B. Funderburg (ed.) Endangered and Threatened Plants and Animals of North Carolina, N.C. State Museum of Natural History, Raleigh.

Constantz, G.D. 1979. Social dynamics and parental care in the tessellated darter (Pisces: Percidae). Proc. Acad. Nat. Sci. Phila. 131:131–138.

Downhower, J.F. & L. Brown. 1977. A sampling technique for benthic fish populations. Copeia 1977:403–406.

Hubbs, C.L. & E.C. Raney. 1946. Endemic fish fauna of Lake Waccamaw, North Carolina. Misc. Publ. Mus. Zool. Univ. Mich. 65: 1–30.

Lindquist, D.G., J.R. Shute & P.W. Shute. 1981. Spawning and nesting behavior of the waccamaw darter, *Etheostoma perlongum.* Env. Biol. Fish. 6:177–191.

Page, L.M. 1983. Handbook of darters. T.F.H. Publications, Neptune City. 271 pp.

Page, L.M. & B.M. Burr. 1976. The life history of the slabrock darter, *Etheostoma smithi,* in Ferguson Creek, Kentucky. Illinois Nat. Hist. Surv. Biol. Notes 99:1–11.

Page, L.M., W.L. Keller & L.E. Cordes. 1981. *Etheostoma (Boleosoma) longimanum* and *E. (Catonotus) obeyense,* two more darters confirmed as egg-clusterers. Trans. Ky. Acad. Sci. 42: 35–36.

Page, L.M., M.E. Retzer & R.A. Stiles. 1982. Spawning behavior in seven species of darters (Pisces: Percidae). Brimleyana 8:135–143.

Shute, P.W., J.R. Shute & D.G. Lindquist. 1982. Age, growth and early life history of the Waccamaw darter. *Etheostoma perlongum.* Copeia 1982:561–567.

Speare, E.P. 1965. Fecundity and egg survival of the central

johnny darter (*Etheostoma nigrum nigrum*) in southern Michigan. Copeia 1965:308–314.
Thresher, R.E. & A.M. Gronnel. 1978. Subcutaneous tagging of small reef fishes. Copeia 1978:352–353.
Winn, H.E. 1956. Egg site selection by three species of darters. Anim. Behav. 5:25–28.

Originally published in Env. Biol. Fish. 11: 107–112

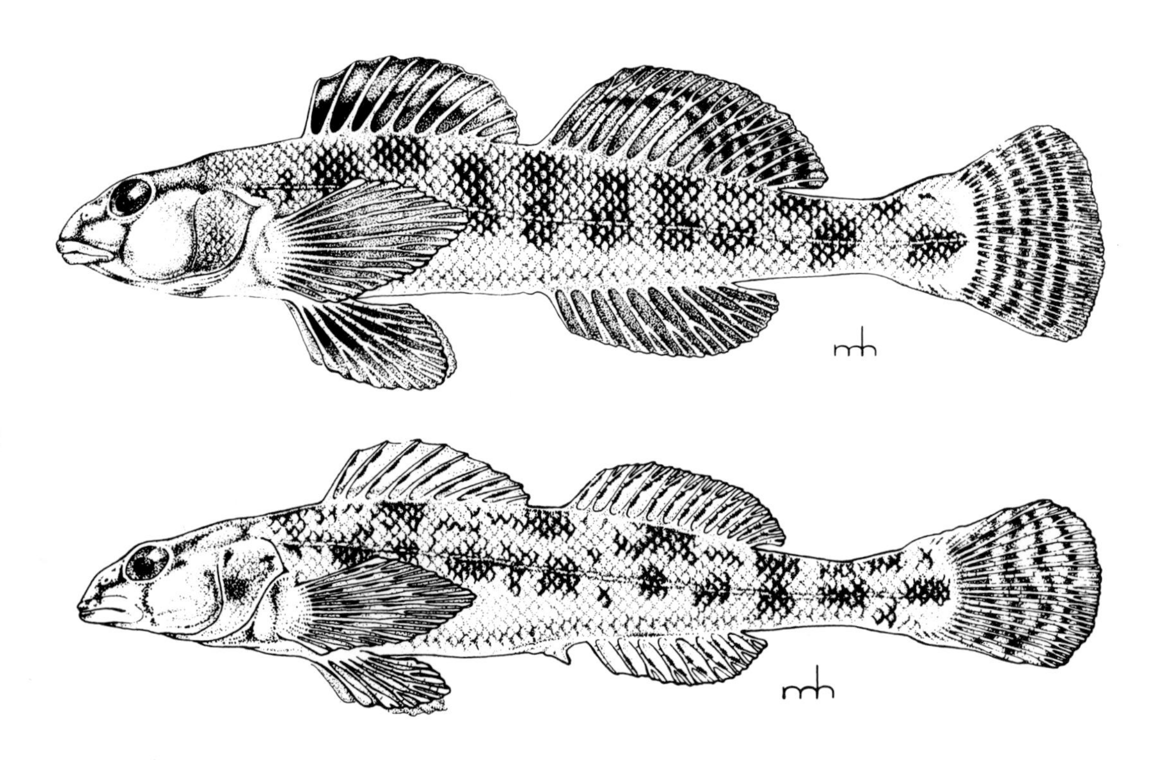

Male (above) and female of the waccamaw darter, *Etheostoma perlongum*.

Diets of four sympatric species of *Etheostoma* (Pisces: Percidae) from southern Indiana: interspecific and intraspecific multiple comparisons

F. Douglas Martin
University of Maryland Center for Environmental and Estuarine Studies, Chesapeake Biological Laboratory, Solomons, MD 20688, U.S.A.

Keywords: Darters, Feeding habits, Stream invertebrates, Order 3 streams, Diet overlap, Stream ecology

Synopsis

The diets of four species of *Etheostoma (E. spectabile, E. caeruleum, E. flabellare* and *E. nigrum)* were investigated from ten Order 3 streams in the White River drainage of southern Indiana. All species fed mainly on insect larvae, primarily chironomids, ephemeropterans and plecopterans. Dietary proportions, as frequency of occurrence, were compared using cluster analysis based on matrices of values of Spearman rank correlation, Schoener index, Jaccard association, and Pearson's r. Certain species and, in some cases, stream pairs within species clustered closely in all analyses or in three of the four. Most clusters do not show closer intraspecific than interspecific similarity and specific streams had more influence than species in forming the logic of some clusters. It can be concluded that these fishes are opportunistic predators.

Introduction

Darter diets attracted attention early as evidenced in studies by Forbes (1880), Pearse (1918) and Turner (1921). This is because there are so many species (126 listed by Robins et al. 1980) and, in part, because they are often the most important consumers of benthic organisms in Order 2 and 3 streams (Small 1975). Because of the numbers of co-occurring species they represent an excellent laboratory for the study of adaptation and competition. Some studies have examined potential competition by comparing diets between species (e.g. Forbes 1880, Thomas 1970, Whitaker 1975, Small 1975, Adamson & Wissing 1977, Cordes & Page 1980) while others have looked at adaptation by examining dietary variation where habitats differ (e.g. Small 1975, Schenck & Whiteside 1977). The focus of the study presented here is to compare both between species and between locations having similar habitats.

Methods and materials

Etheostoma nigrum, E. caeruleum, E. spectabile, and *E. flabellare* were sampled in ten Order 3 streams in the White River drainage (Fig. 1). West Fork White River drainage: Fall Creek, Morgan County, 50 m above Indiana 67 bridge; Burkharts Creek, Morgan County, 50-100 m above Indiana 67 bridge and Buck Creek, Monroe County, Anderson Road, about 0.2 km east of Old 37N. East Fork White River drainage: Leatherwood Creek, Lawrence County at Indiana 58 bridge; Stephens Creek, Monroe County, just upstream from the convergence of Getty Creek; Brummet Creek, Monroe County, 100 m downstream from the convergence of Baby Creek; Lower Schooner Creek, Brown County, 0.4 km downstream from Graveyard Hollow; Jackson Creek, Monroe County, at Indiana 45 bridge; Gulletts Creek, Lawrence County, 100 m upstream from Indiana

David G. Lindquist & Lawrence M. Page (ed.), Environmental biology of darters. ISBN 90-6193-506-7

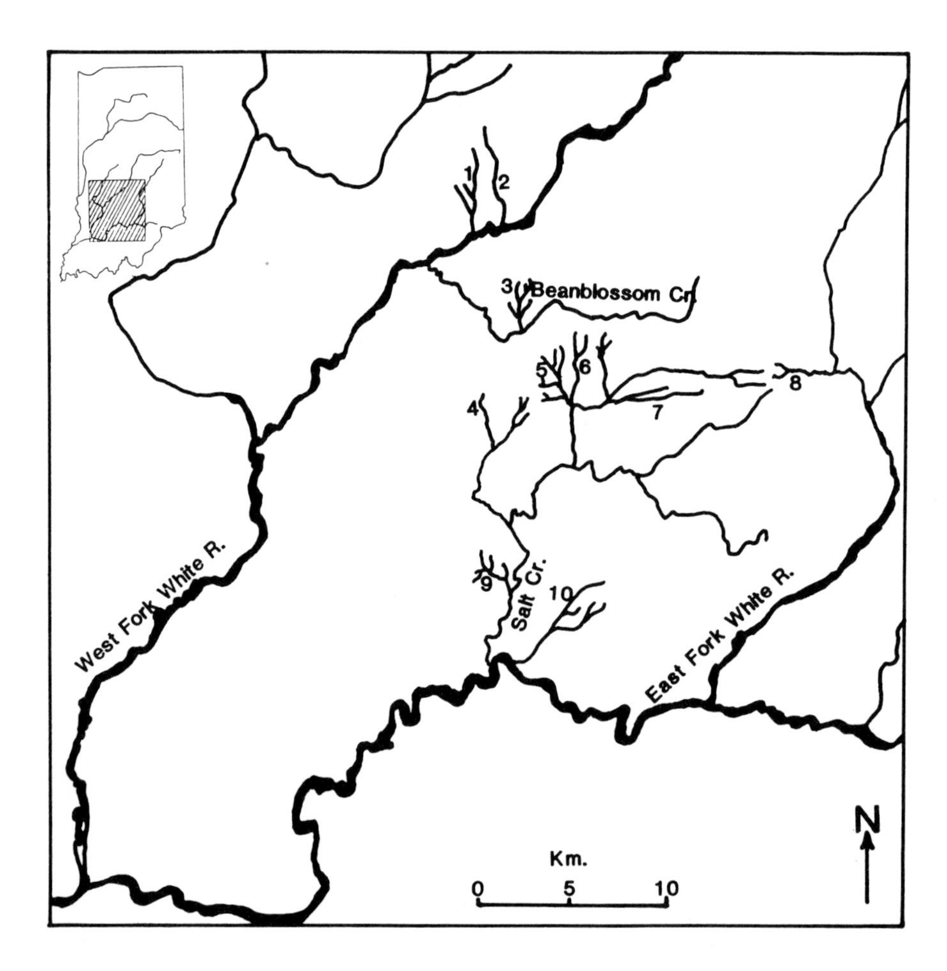

Fig. 1. Collecting localities in Southern Indiana. Creek names corresponding to the numbers are: (1) Fall Creek; (2) Burkhart Creek; (3) Buck Creek; (4) Jackson Creek; (5) Stephens Creek; (6) Brummet Creek; (7) Lower Schooner Creek; (8) Wolf Creek; (9) Gulletts Creek; and (10) Leatherwood Creek.

37 bridge; Wolf Creek, Bartholomew County, at Indiana 46 bridge. All locations were chosen for accessibility, for possessing both pool and riffle habitats, and for having substrates of gravel and cobbles with little silt or sand. Stream Order was determined by the method described by Kuehne (1962). Sampling was done between 15 April and 22 May 1969. Collections were made with 3 × 1 m seines with 6.5 mm stretched mesh. Darters were collected from riffle areas, preserved and stored in 4% formaldehyde.

In the laboratory all darters were identified to species, sexed, and measured to the nearest mm (standard length). The entire digestive tract was excised, placed in a small petri dish with water and the contents removed, identified to lowest practical taxon, and counted.

Formulae used in analysis of the data were the following:

Spearman rank correlation coefficient, corrected for ties, (Siegel 1956)

$$r_s = \frac{\sum x^2 + \sum y^2 - \sum d_i^2}{2\sum x^2 \sum y^2}, \tag{1}$$

where $\sum x^2$ is the sum of ranks for species i, $\sum y^2$ is the sum of squares of ranks for species j and d_i is the difference between ranks of each food category;

$$\text{Schoener index} = 1 - 0.5\left(\sum_{i=1}^{n} |P_{xi} - P_{yi}|\right), \tag{2}$$

where P_{xi} is the proportion of food category in the diet of species x, P_{yi} is the proportion of food category in the diet of species y, and n is the number of food categories; Jaccard coefficient of association,

$$J = \frac{c}{a + b - c} \times 100 , \quad (3)$$

where a is the total number of food categories in the diet of species x, b is the total number of food categories in the diet of species y and c is the number of food categories common to both species; and Pearson's r,

$$r = \frac{\sum^{i}(x_i - \bar{x})(y_i - \bar{y})}{\sqrt{\sum^{i}(x_i - \bar{x})^2 \sum^{i}(y_i - \bar{y})^2}} ,$$

where x_i is the frequency of food category i in the diet of species 1 and y_i is the frequency of food category i in the diet of species 2. So that all four methods of analysis would be directly comparable, frequency of occurrence was used for all. After indices were calculated they were clustered using an average distance procedure.

Only food habit data from sexually mature individuals were used in these analyses. F tests of all within-stream, within-species comparisons based on 5 mm size increments showed no significant differences. Likewise, Spearman rank correlation coefficient values indicated significant similarity, and all such comparisons using the Schoener index have values above 0.65. Schoener index values above 0.60 are usually considered to be 'biologically significant' (Zaret & Rand 1971, Mathur 1977, Wallace 1981). Consequently, all size categories were considered equivalent, and the data were lumped. Similar results were obtained between sexes, and data for the sexes were lumped. *Etheostoma nigrum* and *E. flabellare* had sample sizes of less than 8 from each creek, samples for each of these two species were lumped.

Results and discussion

Table 1 summarizes per cent occurrence of food items for the total sample. Algae are listed, although these occurrences probably represented accidental ingestion while feeding on invertebrates and contribute little nutrition.

Figures 2-5 are dendrograms generated using the Spearman rank correlation coefficient, Schoener index, Jaccard index of association, and Pearson's r, respectively. Table 2 lists all pairs of populations which cluster closely in three or four of these dendrograms. Of the seven pairs in Table 2, four involved *E. spectabile* from Stephens Creek and three involved *E. spectabile* from Buck Creek. Only two are interspecific pairs. There are 105 possible pairwise combinations of which 48 are intraspecific pairs and 57 are interspecific pairs. A Chi square test with the assumption that the expected proportion would be the same as the proportion of possible combinations indicated that intraspecific pairs are significantly more frequent ($\chi^2 = 6.48$, df = 1, $p<0.02$) than expected by random chance. Six of the seven pairs were some combination of Buck Creek, Stephens Creek, Brummet Creek or Lower Schooner. The last three of these are all part of the Salt Creek subdrainage of the East Fork White River drainage; Buck is part of the West Fork White River drainage. Despite being parts of different drainages these creeks are all located close together, drain very similar geological formations and have extremely similar morphologies as far as stream width, depth and bottom type. However, physical measurements were not made, and it is not clear that these streams are more similar to one another than are other combinations.

This study suggests, as have most other studies, that darters do not partition the habitat simply on the basis of food resources (Braasch & Smith 1967). The few studies which implicate partitioning among darters do so on the basis of foraging area (Smart & Gee 1979), habitat (Page & Schemske 1978) or stream type (Kuehne 1962). Nonpartitioning on the basis of food is not surprising inasmuch as Small (1975) noted that in Order 2 and 3 streams in his study area the only fish feeding on benthic invertebrates are darters, sculpins, and minnows of the genus *Rhinichthys*. The latter two groups were not present in my collections and there is probably a large resource base on which the darters feed.

Wallace (1981) discusses reasons why per cent occurrence is often a poor measures of diet over-

Table 1. Per cent occurrence of food items summarized for all creeks. Numbers in parentheses are numbers of darters examined.

Food items	*E. spectabile* (371)	*E. caeruleum* (76)	*E. nigrum* (20)	*E. flabellare* (22)
Algae	0.5	–	–	–
Gastropoda	–	1.3	–	–
Oligochaeta	1.6	–	–	–
Copepoda	0.5	–	–	–
Ostracoda	2.4	1.3	5.0	–
Amphipoda	4.8	–	–	–
Isopoda	11.8	6.6	5.0	13.6
Decapoda	–	2.6	–	–
Insecta:				
Odonata	0.8	–	5.0	–
Ephemeroptera	40.7	50.0	30.0	77.3
Plecoptera	28.6	6.6	–	–
Hemiptera	1.6	5.3	–	–
Neuroptera	–	–	5.0	–
Trichoptera	8.1	13.2	10.0	13.6
Coleoptera	0.3	1.3	–	–
Diptera:				
Tipulidae	1.1	1.3	–	–
Simuliidae	12.7	9.2	5.0	13.6
Chironomidae:				
Chirominae	10.2	28.9	30.0	22.7
Orthocladiinae	49.3	44.7	55.0	22.7
Tanypodinae	3.2	9.2	–	4.5
Cyclorhapha	6.5	1.3	15.0	–
Fish eggs	19.1	9.2	5.0	4.5

Table 2. Pairs of populations (parenthetical numbers refer to locations in Fig. 1) and the number of times that they clustered closely. Close clustering is defined as amalgamation distances less than the median for the points of amalgamation. These medians are: Pearson's r = 0.85; Spearman rank = 0.72; Jaccard = 0.67; Schoener Index = 0.69. (Maximum co-occurrence = 4).

Pair	*No. of times clustered*
E. spectabile (3) - *E. spectabile* (5)	4
E. spectabile (3) - *E. spectabile* (6)	3
E. spectabile (3) - *E. spectabile* (7)	3
E. spectabile (5) - *E. spectabile* (6)	3
E. spectabile (5) - *E. spectabile* (7)	3
E. spectabile (5) - *E. caeruleum* (6)	3
E. caeruleum (9) - *E. nigrum*	3

lap; however, it is a more robust measure at low sample size and because this study was not an attempt to numerically compare species on an energetic basis, it was deemed adequate. The Jaccard association, used because the cluster analysis procedure suggested by Bortone et al. (1981) gave good separation at species levels when comparing multiple species and locations, further limited the type of data because it uses only presence or absence data.

Each of the analytic techniques has limitations. The Jaccard index of association is sensitive to a combination of rare food items and uneven sampling effort. In a simple system with four food categories and complete overlap, the addition of a single item in a fifth category and in only one individual of the sample will change the value 100 to 80. With the exceptions of the *E. spectabile* samples from Lower Schooner Creek, Fall Creek and Gulletts Creek both the numbers of stomachs and the numbers of food categories are high enough (Table 3) that single individuals will have small impacts on the values. The Spearman rank correlation coefficient suffers from the same weakness concerning rare food items (Wallace 1981) but is less sensitive to sample size.

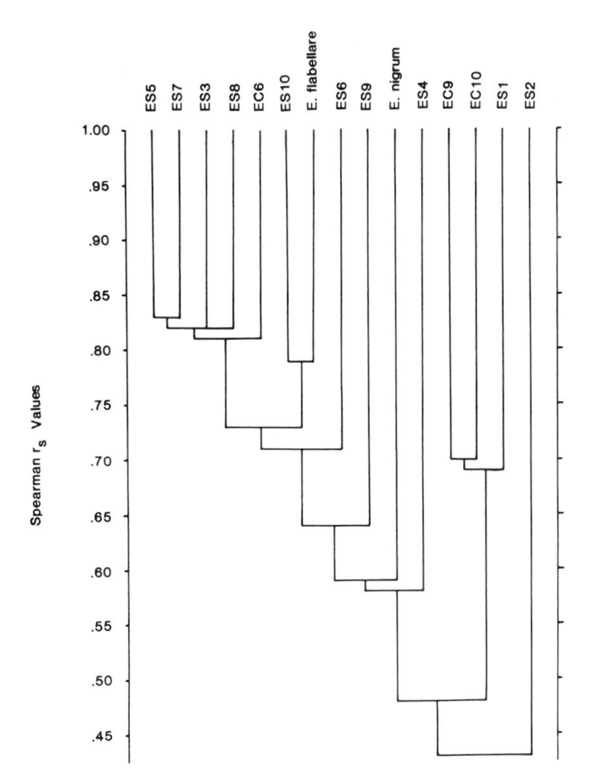

Fig. 2. Dendrogram of similarities in diet based on Spearman rank correlation coefficient values. Minimum and maximum values are −1 and +1 respectively. Es refers to *E. spectabile,* Ec refers to *E. caeruleum.* The numbers refer to creeks sampled (see Fig. 1.)

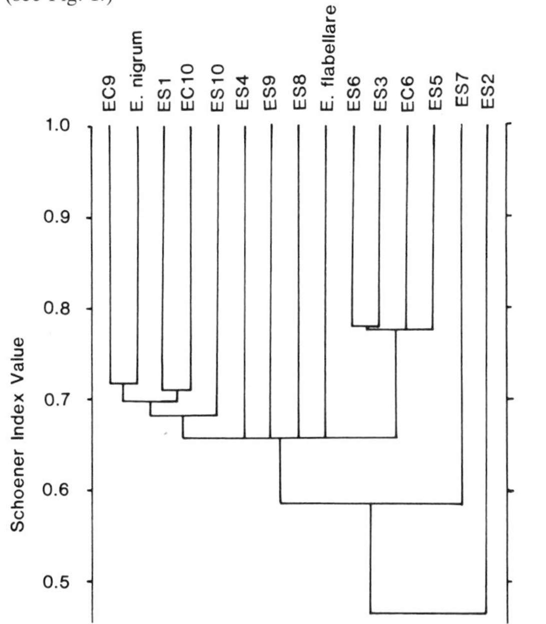

Fig. 3. Dendrogram of similarities in diet based on Schoener (1968) index values. 0 = no overlap; 1 = complete overlap. Symbols and numerals explained in figure 2.

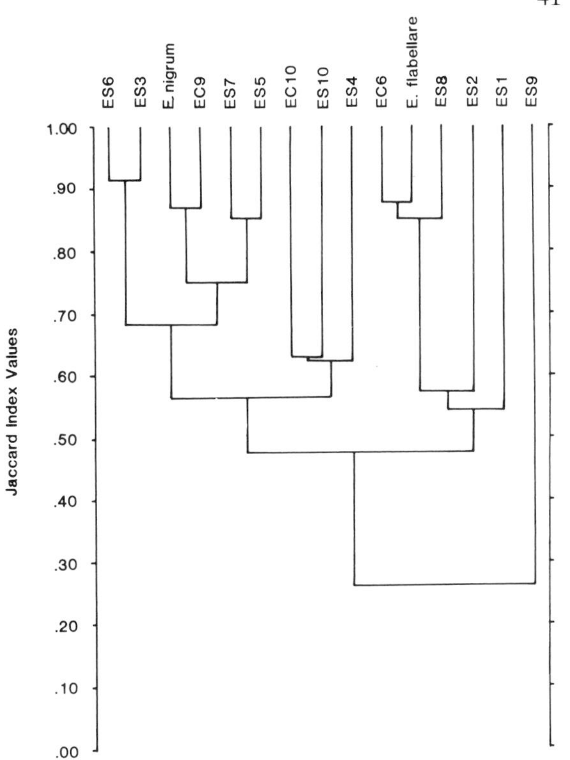

Fig. 4. Dendrogram of similarities in diet based on Jaccard (1902) *fide* Udvardy (1969) index of association values. 0 = no overlap, 1.00 = complete overlap. See figure 2 for explanation of other symbols.

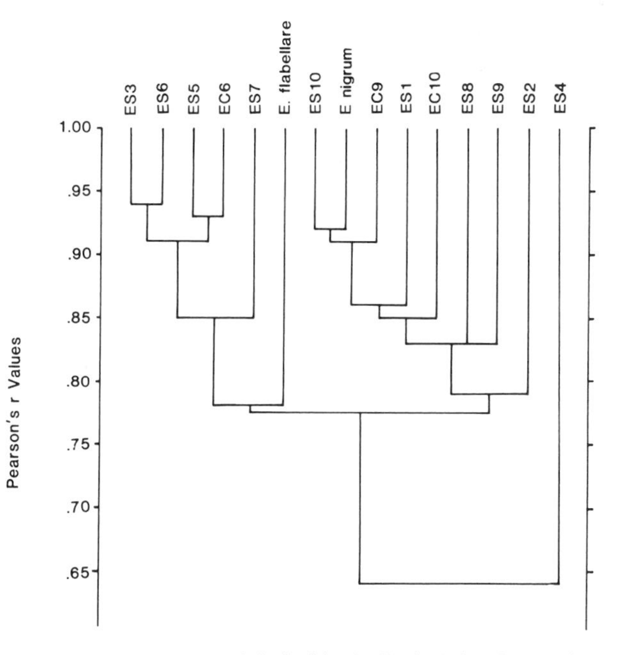

Fig. 5. Dendrogram of similarities in diet based on Pearson's r values. −1 = perfect negative correlation, +1 = perfect positive correlation.

Table 3. Sample sizes for this study. Parenthetical numbers refer to locations in Figure 1.

Species	Stream	n	No. food categories
E. spectabile	Fall Creek (1)	7	6
	Burkhart Creek (2)	21	10
	Buck Creek (3)	67	13
	Leatherwood Creek (10)	14	7
	Stephens Creek (5)	61	15
	Brummet Creek (6)	122	13
	Lower Schooner Creek (7)	8	6
	Jackson Creek (4)	36	10
	Gulletts Creek (9)	6	6
	Wolf Creek (8)	29	11
E. caeruleum	Leatherwood Creek (10)	9	6
	Brummet Creek (6)	29	8
	Gulletts Creek (9)	38	13
E. nigrum	All creeks	20	11
E. flabellare	All creeks	22	8

Table 4. Comparison of the four dendrograms using Mantel's (1967) method for comparing matrices. The upper number is the Z score which is the observed association between cells of the test matrices. The lower number is the t score. An asterisk by the t score indicates significant similarity ($\alpha = 0.05$).

	Spearman's r_s	Schoener Index	Jacard Index
Schoener Index	74.84 −0.03		
Jaccard Index	63.79 0.88	69.09 *3.44	
Pearson's r	94.39 −0.11	101.99 0.01	84.75 0.17

Wallace (1981) concluded that the Schoener index is the least objectionable of the indices available when resource-availability data are not present. Linton et al. (1981) reported that, for values below 0.85, this index gave the closest estimate of real overlap of the four indices that they compared. It is coupled here with a cluster analysis technique to aid in visualizing results.

To use Pearson's r, the assumptions are that sample sizes are adequate to represent all common food categories, and that the variations are normally distributed. It has an advantage over rank correlation coefficients in that the difference between a category with few items and a category having only one item more makes only a slight difference in the calculated linear coefficient.

Since each of these indices has major problems that prevent uncritical use, I felt that use of several indices followed by cluster analysis of the index values would provide the most biologically meaningful analysis of the data. Methods of comparing dendrograms are not well developed (Strauss 1982) but if dendrograms are transformed to matrices of amalgamation distances they can be compared using the Mantel (1967) test (Table 4). The Z scores are all similar so that they are all similarly associated as pairs; however, only the matrices of Jaccard and Schoener values had t values indicating significant similarity. This should be interpreted cautiously as the two indices are calculated on different aspects of the data and have no logical reason why they should necessarily be associated.

All analytic methods used test different aspects of the data set. From the low number of population pairs which cluster together consistently it seems that these species are all foraging opportunistically in similar subareas of the riffles which is consistent with results of Adamson & Wissing

Table 5. Darters reported to feed on fish eggs. Where choice is available spring data only are presented.

Species	Source	Occurrence	n
E. barbouri	Flynn & Hoyt 1979	~ 12[1]	18
E. caeruleum	Adamson & Wissing 1977	2.6[2]	82
E. caeruleum	This study	9.2[1]	76
E. flabellare	Adamson & Wissing 1977	1.8[2]	62
E. flabellare	This study	4.5[1]	22
E. nigrum	This study	5.0[1]	20
E. radiosum cyanorum	Scalet 1972	0.2[3]	600
E. radiosum cyanorum	Scalet 1974	0.3[1]	600
E. spectabile	This study	19.1[1]	371
P. maculata	Thomas 1970	4.5[1]	22
P. shumardi	Thomas 1970	8.0[1]	12

[1] Reported as % frequency of occurrence.
[2] Reported as % total dry weight.
[3] Reported as % total numbers of food items.

(1977). While opportunistic feeding is suggested, the fact that most of the clusters which were consistent among methods involved only *E. spectabile* suggested that this species has a feeding niche separate from the other three species. Partitioning on habitat can not be ruled out though as the creeks which were involved had *E. spectabile* as the sole riffle species or had only *E. spectabile* and a few *E. caeruleum* individuals.

All four species feed heavily on larvae of chironomids, plecopterans and ephemeropterans. Forbes (1880) noted that species with either very small adult body size, such as *E. microperca*, or larger than average adult body size for darters, such as *Percina caprodes*, tend to feed on crustaceans while those with average adult body size feed primarily on larval insects. These observations have been verified by Burr & Page (1978, 1979), Schenck & Whiteside (1977), and Cordes & Page (1980). The species of darters studied here are of a average size and should be expected to have the diet typical of average sized darters.

Darters rely on vision for food location (Roberts & Winn 1962, Mathur, 1973, Schenck & Whiteside 1977) and with few exceptions (Smart & Gee 1979) feed on benthic organisms (Fahy 1954, Small 1975). Since movement is usually required to attract a darter to a food item (Roberts & Winn 1962) it is surprising how frequently fish eggs are reported in their diets. Table 5 lists species which are reported to prey on eggs. *E. spectabile* from this study consumed more eggs than has been reported for other species. Because of the method of reporting data the two species reported by Adamson & Wissing (1977) can not be directly compared, but *E. caeruleum* may approach this frequency.

Acknowledgements

This work was supported by contract 342–303.721 from the Indiana Department of Natural Resources. Computer time was furnished by the University of Maryland Computer Center. Mary Hengstebeck helped to analyze these data. I thank Dennis Bertram and Paul S. Choitz for their contributions. Contribution Number 1414, Center for Environmental and Estuarine Studies of the University of Maryland.

References cited

Adamson, S.W. & T.E. Wissing, 1977. Food habits and feeding periodicity of the rainbow, fantail and banded darters in Four Mile Creek. Ohio J. Sci. 77: 164–169.

Bortone, S.A., D. Siegel & J.L. Oglesby. 1981. The use of cluster analysis in comparing multi-source feeding studies. North-east Gulf Sci. 5: 81–86.

Braasch, M.E. & P.E. Smith. 1967. The life history of the slough darter, *Etheostoma gracile* (Pisces, Percidae). Ill. Nat. Hist. Surv. Biol. Notes 58: 1–12.

Burr, B.M. & L.M. Page. Life history of the cypress darter,

Etheostoma proeliare, in Max Creek, Illinois. Ill. Nat. Hist. Surv. Biol. Notes 106: 1–15.

Burr, B.M. & L.M. Page. 1979. The life history of the least darter, *Etheostoma microperca*, in the Iroquois River. Ill. Nat. Hist. Surv. Biol. Notes 112: 1–15.

Cordes, L.E. & L.M. Page. 1980. Feeding chronology and diet composition of two darters (Percidae) in the Iroquois River System, Illinois. Amer. Midl. Nat. 104: 202–206.

Fahy, W.E. 1954. The life history of the northern greenside darter, *Etheostoma blennioides blennioides* Rafinesque. J. Elisha Mitchell Sci. Soc. 70: 139–205.

Flynn, R.B. & R.D. Hoyt. 1979. The life history of the teardrop darter, *Etheostoma barbouri* Kuehne and Small. Amer. Midl. Nat. 101: 127–141.

Forbes, S.A. 1880. The food of the darters. Amer. Nat. 14: 697–703.

Jaccard, P. 1902. Lois de distribution florale dans la zone alpine. Bull. Soc. Vandoise Sci. Nat. 38: 69–130.

Kuehne, R.A. 1962. A classification of streams, illustrated by fish distribution in an eastern Kentucky creek. Ecology 43: 608–614.

Linton, L.R., R.W. Davies & F.J. Wrona. 1981. Resource utilization indices: an assessment. J. Anim. Ecol. 50: 283–292.

Mantel, N. 1967. The detection of disease clustering and a generalized regression approach. Cancer Research 27: 209–220.

Mathur, D. 1973. Food habits and feeding chronology of the blackbanded darter, *Percina nigrofasciata* (Agassiz), in Halawakee Creek, Alabama. Trans. Amer. Fish. Soc. 102: 48–55.

Mathur, D. 1977. Food habits and competitive relationships of the bandfin shiner in Halawakee Creek, Alabama. Amer. Midl. Nat. 97: 89–100.

Page, L.M. & D.W. Schemske. 1978. The effects of interspecific competition on the distribution and size of darters of the subgenus *Catonotus* (Percidae: *Etheostoma*). Copeia 1978: 406–412.

Pearse, A.S. 1918. The food of the shore fishes of certain Wisconsin lakes. Bull. US. Bur. Fish. 35: 249–292.

Roberts, N.J. & H.E. Winn. 1962. Utilization of the senses in feeding behavior of the johnny darter, *Etheostoma nigrum*. Copeia 1962: 567–570.

Robins, C.R., R.M. Bailey, C.E. Bond, J.R. Brooker, E.A. Lachner, R.N. Lea & W.B. Scott. 1980. A list of common and scientific names of fishes from the United States and Canada (4th edition). Amer. Fish. Soc. Special Publ. No. 12, Bethesda. 174 pp.

Scalet, C.G. 1972. Food habits of the orangebelly darter, *Etheostoma radiosum cyanorum* (Osteichthys: Percidae). Amer. Midl. Nat. 87: 515–522.

Scalet, C.G. 1974. Lack of piscine predation of the orangebelly darter, *Etheostoma radiosum cyanorum*. Amer. Midl. Nat. 92: 510–512.

Schenk, J.R. & B.G. Whiteside, 1977. Food habits and feeding behavior of the fountain darter, *Etheostoma fonticola* (Osteichthys: Percidae). Southwest Nat. 21: 487–492.

Schoener, T.W. 1968. The *Anolis* lizards of Bimini: resource partitioning in a complex fauna. Ecology 49: 704–726.

Siegel, S. 1956. Nonparametric statistics for the behavioral sciences. McGraw-Hill, New York. 312 pp.

Small, J.W. Jr. 1975. Energy dynamics of benthic fishes in a small Kentucky stream. Ecology 56: 827–840.

Smart, H.J. & J.H. Gee. 1979. Coexistence and resource partitioning in two species of darters (Percidae), *Etheostoma nigrum* and *Percina maculata*. Can. J. Zool. 57: 2061–2071.

Strauss, R.E. 1982. Statistical significance of species clusters in association analysis. Ecology 63: 634–639.

Thomas, D.L. 1970. An ecological study of four darters of the genus *Percina* (Percidae) in the Kaskaskia River, Illinois. Ill. Nat. Hist. Surv. Biol. Notes 70: 1–18.

Turner, C.L. 1921. Food of the common Ohio darters. Ohio J. Sci. 22: 41–62.

Udvardy, M.D.F. 1969. Dynamic zoogeography with special reference to land animals. Van Nostrand Reinhold Co, New York. 445 pp.

Wallace, R.K. Jr. 1981. An assessment of diet-overlap indexes. Trans. Amer. Fish. Soc. 110: 72–76.

Whitaker, J.O., Jr. 1975. Foods of some fishes from the White River at Petersburg, Indiana. Proc. Ind. Acad. Sci. 84: 491–499.

Zaret, T.M. & A.S. Rand. 1971. Competition in tropical stream fishes: support for the competitive exclusion principle. Ecology 52: 336–342.

Originally published in Env. Biol. Fish. 11: 113–120

Life history of the gulf darter, *Etheostoma swaini* (Pisces: Percidae)

David L. Ruple[1], Robert H. McMichael, Jr.[2] & John A. Baker[3]
[1] *Gulf Coast Research Laboratory, East Beach Drive, Ocean Springs, MS 39564, U.S.A.*
[2] *Marine Research Laboratory, Department of Natural Resources, 100 8th Ave. S.E., St. Petersburg, FL 33701, U.S.A.*
[3] *Environmental Laboratory, U.S. Army Engineer Waterways Experiment Station, Vicksburg, MS 39180, U.S.A.*

Keywords: Ecology, Habitat, Feeding, Age, Growth, Reproduction, Mississippi

Synopsis

Etheostoma swaini, the gulf darter, was collected from the Black Creek drainage in southern Mississippi (February 1978 — April 1979). The gulf darter generally inhabits small- to moderate-size creeks and occurs over a sand or sandy mud bottom, often in association with aquatic vegetation or a layer of organic debris. Larval dipterans were the most important food items, both numerically and volumetrically. Chironomids were found in 71–100% of the stomachs in all except the unusual March 16 collection. The length frequency distribution and the scale annuli analysis indicated there were three year-classes present in the population at any one time. Fifty-one percent of the specimens taken were less than 12 months old. During the mid-February to late March spawning season gulf darters were most often collected over clean gravel or gravel-sand substrates. Laboratory observations suggest that the female burrows into the gravel where the demersal, adhesive eggs are deposited. Female gulf darters significantly outnumbered males at a ratio of 59:41.

Introduction

Etheostoma swaini (Jordan), the gulf darter, is found in small- to medium-size sandy, gravelly, vegetated streams of the Gulf slope drainages from Lake Pontchartrain, Louisiana east to the Ochlockonee River drainage, Florida, and in many eastern tributaries to the Mississippi River from Buffalo Bayou, Mississippi north to the Obion River system of Tennessee and Kentucky (Starnes 1980). *Etheostoma swaini* is reported to be one of only a limited number of *Etheostoma* species to be typically found in lowland swamplike stream habitats of the Gulf of Mexico drainage. Fishes of the genus *Etheostoma* are most often restricted to streams of comparatively high oxygen levels and low temperature (Ultsch et al. 1978).

Etheostoma swaini has been placed in the *E. asprigene* species group within the subgenus *Oligocephalus*, along with *E. ditrema* and *E. nuchale* (Page 1981). Ramsey & Suttkus (1965) believed *E. swaini* to be most closely related to *E. hopkinsi*, while Howell & Caldwell (1965) discussed the close relation to *E. nuchale*. Nothing has been previously published on the life history of *E. swaini* other than its generalized habitat.

This paper reports on various aspects of the life history of *E. swaini*, including food habits, reproduction, species association, growth, and preferred habitats, in the Pascagoula River drainage of southern Mississippi.

David G. Lindquist & Lawrence M. Page (ed.), Environmental biology of darters. ISBN 90-6193-506-7

Study area

Etheostoma swaini were collected from the Black Creek system in southern Mississippi (Fig. 1). Black Creek arises in Jefferson Davis County and flows southeast approximately 120 km to its confluence with the Pascagoula River in Jackson County. The study area was located in Forrest and Lamar counties, northwest of U.S. Route 49. Black Creek, Black Tom Creek, and Perkins Creek were the principal bodies of water sampled. A few collections were also made in Sandy Run, Boggy Hollow, and Mixon Creek.

Streams within the system are spring-fed and relatively clear, except during periods of heavy rainfall. The presence of dissolved tannic acid from decaying humus often gives the water a brownish tinge. The pH ranged from 5.2 to 6.8 (Baker & Ross 1981), reflecting the low soil pH. Streams within the study area ranged from 1 to 20 m wide. Gradients were from over 6 m km^{-1} in some of the tributaries to 1.1 m km^{-1} in most of Black Creek.

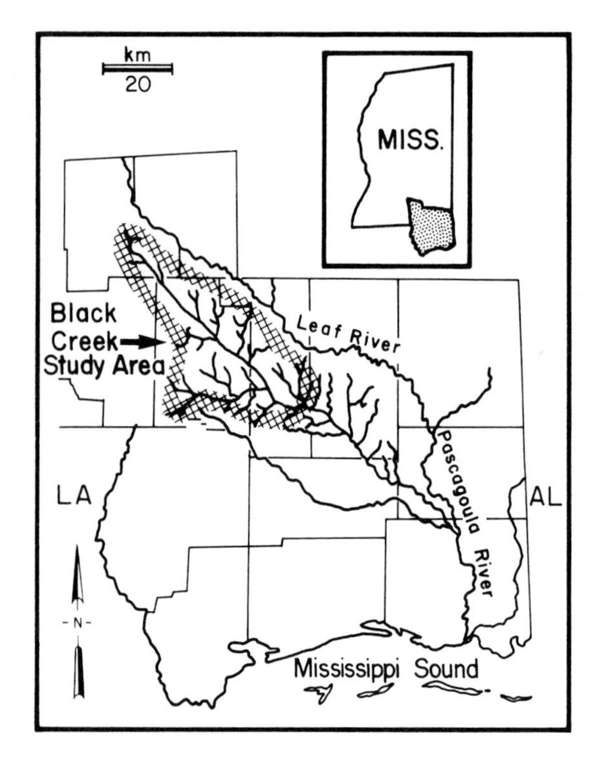

Fig. 1. A drainage map of the Black Creek, Mississippi study area (from Baker & Ross 1981).

Materials and methods

Monthly collections of *E. swaini* were taken from the Black Creek drainage between February 1978 and April 1979. No collections were taken during November due to high water. A total of 540 specimens was collected, fixed in 10% formalin, and later transferred to 45% isopropanol. Most collections were made with a 2.0 × 1.2 m, 3.1 mm mesh seine. Occasionally a larger seine, 3.6 × 1.2 m, 3.1 mm was used, particularly in the wider sections of Black Creek and Mixon Creek.

Several physical parameters were recorded at each locality to determine darter microhabitat preference. We defined microhabitat as an area within a stream which has a unique combination of physical characteristics, and which can be readily differentiated from other microhabitats on the basis of these characteristics. We determined the surface current velocity by timing a partially submerged object over a known distance. We measured maximum and minimum water depths to the nearest cm, mean depths were calculated from approximately 8 depth measurements taken over the entire microhabitat. We also determined amount and type of bottom cover, aquatic vegetation, litter covering the substrate, and the type of substrate. Air and water temperature were recorded at each station. Underwater visual observations were made in the field to confirm microhabitat preferences and species associations.

Specimens were blotted to remove excess water and weighed to the nearest 0.01 g on an electric, single pan balance. Total (TL) and standard (SL) lengths were measured using methods described by Hubbs & Lagler (1964).

Ages of darters were determined from analysis of length-frequency data and by examining scale annuli. Annuli were counted on scales removed from the left side of the body, below the lateral line and posterior to the pectoral fin. Scales were wet mounted and viewed under a dissecting microscope (25 ×). A standard length to total body weight relationship was determined using the formula: $\text{Log } W = \text{Log } a + n \text{ Log } L$, where W equals whole body weight (g) and L is standard length (mm), and a and n are constants.

Stomach contents were identified and enumerated for at least 10 specimens per month. Most stomachs examined were from collections made between 1200 and 1600 h. Food habit analysis was based on the percentage of stomachs in which food item categories occurred, and on the mean number of items of each category per stomach. The relative volumetric contribution of each food item category was estimated by eye but actual volumes were not quantified. Stomach fullness for each fish, compared to a similar sized fish with a packed stomach, was visually estimated as: 1 = 1–25% full; 2 = 26–50% full; 3 = 51–75% full; and 4 = >75% full.

Sex was determined by a visual observation of gonads. Gonads were blotted and weighed to the nearest 0.0001 g on an electric, single pan balance. Gonadal Somatic Indices

$$GSI = \frac{\text{ovary weight}}{\text{whole body weight}}$$

were determined for both males and females. Mean monthly GSI values were calculated to show the degree of reproductive readiness, assuming that gonadal development (gonad weight) reached a maximum just prior to spawning. Gonad development was determined for 133 males and 195 females. Mature ova (large with yellow yolk) were counted (total fecundity) from randomly selected specimens taken during February and March 1978. Relative fecundity was defined as number of mature ova per gram, ovary-free body weight, i.e.

$$\text{relative fecundity} = \frac{\text{total fecundity}}{\text{whole body weight} - \text{ovary weight}} .$$

Results and discussion

Habitat preference and species associations

Etheostoma swaini inhabits small- to moderate-size creeks (1–20 m wide) throughout the Black Creek drainage. Gulf darters were usually collected over a sand or sandy mud bottom, often with a considerable layer of organic debris. During the February to March spawning season, however, gulf darters were found most often over clean gravel or gravel-sand substrates. Mean current speed of gulf darter microhabitats also differed significantly ($t = 1.87$, $df = 23$, $p < 0.05$) between the spawning season ($\bar{x} = 47.4$ cm s^{-1}) and the remainder of the year ($\bar{x} = 33.8$ cm s^{-1}). Recently hatched darters probably inhabited areas of reduced current speeds, however, by ~13–15 mm SL they were utilizing the adult microhabitat.

Gulf darters frequented extremely shallow locations, often foraging in water less than 5 cm deep. Mean and maximum depths of microhabitats in which we collected *E. swaini* were 32 and 46 cm, respectively, and showed no seasonal differences. Both juvenile and adult darters utilized similar depths.

In most creeks the microhabitat of *E. swaini* was characterized by moderate to heavy amounts of aquatic vegetation, primarily *Sparganium americanum. Orontium aquaticum, Mayaca fluviatilis,* and *Eleocharis* spp. were also utilized, but less often than *Sparganium*. Gulf darters actively foraged in and among clumps of vegetation and they also utilized the areas of reduced current downstream from these clumps as resting sites. In larger, predominantly unvegetated creeks *E. swaini* inhabited quiet streamside areas of sand and sand-silt substrates. In such microhabitats they were considerably less abundant than *Etheostoma stigmaeum, Ammocrypta beani* or *Percina nigrofasciata*.

The above microhabitat characterization for gulf darters during February, 1978 to April, 1979 corresponds well with data collected by S.T. Ross (personal communication) for a wider area from 1975 to 1981. The description also agrees with that reported by Swift et al. (1977) for the Ochlockonee River, Florida and Georgia. In the smallest creeks, however, *E. swaini* may occasionally be found in shallow, swift riffles formed by logs, rocks or vegetation. Such localities may have formed the basis for the brief habitat description given by Douglas (1974).

In its preferred microhabitat, *E. swaini* was usually associated with *Noturus leptacanthus* and *P. nigrofasciata*, and often with *Ichthyomyzon gagei* larvae. *Percina nigrofasciata* appeared to be the most ecologically similar species to *E. swaini*; however, the extent to which they may compete is not known. *Percina nigrofasciata* foraged in a much wider variety of microhabitats and did not exploit vegetation or organic debris to the extent the gulf darter did. In areas of still water, *P. nigrofasciata* was occasionally observed to forage several cm above the substrate. Gulf darters were entirely restricted to the substrate. *Noturus leptacanthus* forages exclusively after dark (Clark 1978), and lamprey larvae are deposit feeders. Thus, competition for food between *E. swaini* and these species is unlikely.

In sparsely vegetated, clean substrate microhabitats gulf darters were associated with *Notropis texanus, N. roseipinnis* and *E. stigmaeum*. *Notropis texanus* generally frequents such areas (Baker & Ross 1981) and is not restricted to substrate foraging, while *N. roseipinnis* feeds in the upper water column (Baker & Ross 1981). *Etheostoma stigmaeum* appears to usually inhabit more sluggish, unvegetated habitats (S.T. Ross personal communication).

No agonistic behavior toward other species was noted. In many instances, *E. swaini* and *P. nigrofasciata* were observed foraging nearly side-by-side for several minutes.

Food habits

A total of 222 stomachs were examined from 15 different collections. Nineteen categories of food items were identified, with the majority representing aquatic invertebrates (Table 1). No dietary differences occurred between sexes, and with the exceptions noted below, few differences were found among size classes. Overall, the diet of gulf darters was similar to that found for other *Etheostoma* species (Burr & Page 1978, Braasch & Smith 1967, Flynn & Hoyt 1979, Karr 1964, Page 1974, 1975, 1980, Page & Burr 1976).

Larval dipterans were the most numerous food item in fish of all sizes. Chironomids were found in 71–100% of the stomachs examined each month, except in the unusual March 16 collection when they were absent. Simulids also occurred in many stomachs but were seldom numerous. Overall, dipterans comprised the bulk of the biomass consumed by all sizes of darters. There were no significant quantitative differences in the utilization of chironomids by size classes of gulf darters. Adults and young-of-year fish both consumed similar numbers of chironomids. However, young-of-year fish ate primarily small chironomid species, and early instars of larger species, while adult fish consumed a wide size range of chironomids.

Ephemeroptera nymphs were most regularly consumed by gulf darters during spring and early summer. All but the very smallest darters utilized mayflies (Heptageniidae, *Stenonema*) to some extent. Ephemeroptera nymphs were judged to be second in overall volumetric importance.

Trichoptera larvae occurred regularly in darter stomachs throughout much of the year. Caddisflies consumed were almost entirely final instar (case-dwelling) *Hydroptila* spp. (family Hydroptilidae). These very small caddisflies were seldom numerous in individual stomachs and comprised only a small percentage of the diet volumetrically. Rarely, larger caddisfly species (Hydropsychidae, *Cheumatopsyche* spp.) were encountered. All sizes of gulf darters utilized caddisflies to some extent.

Microcrustaceans (Cladocera, Copepoda, Ostracoda) and Acarina were regular components of gulf darters diets, but they were relatively insignificant volumetrically. Young-of-year darters ate greater numbers of microcrustaceans than did older fish during late spring and early summer, when the size difference between year classes was greatest. On May 23, young-of-year fish (9.9–18.4 mm SL) had an average of 15.13 microcrustaceans per stomach while adult fish (31.8–42.5 mm SL) averaged only 1.40 per stomach. The linear regression of organism number with SL was highly significant ($r = -0.73$, $n = 18$, $p < 0.001$). This relationship was still evident until June 27 ($r = -0.48$, $n = 29$, $p < 0.01$), though weaker. Subsequent collections, in which young-of-year darters had grown to 18.9–25.8 mm SL, showed no significant correlation between the numbers of these food items and the size of darters.

Isopods and Anisoptera nymphs were the most important of the remaining food item categories. Isopods comprised an estimated 50% of the combined volume of all darter stomachs during the

Table 1. Stomach contents of *Etheostoma swaini* from the Black Creek drainage, Mississippi, from February 8, 1978 to April 17, 1979. Data are presented as percentage of stomachs in which each food category occurred, with mean number of food items per stomach below in parentheses. Numbers below dates indicate numbers of stomachs examined.

Food organism category	Feb.8 (10)	Mar.4 (10)	Mar.16 (10)	Apr.27 (11)	May23 (18)	Jun.15 (11)	Jun.27 (29)	Jul.26 (17)	Aug.16 (19)	Sep.24 (13)	Oct.25 (13)	Dec.6 (14)	Jan.29 (12)	Feb.28 (17)	Apr.17 (18)
Oligochaeta				18 (0.18)			3 (0.03)							6 (0.12)	
Gastropoda															
Ancylidae				9 (0.09)							8 (0.08)			12 (0.12)	
Arachnida															
Acarina	10 (0.10)			45 (2.18)	39 (0.89)	27 (0.27)	21 (0.38)	12 (0.71)	5 (0.05)	15 (0.23)	8 (0.08)	21 (0.21)		18 (0.24)	
Crustacea															
Cladocera	70 (2.90)	10 (0.10)		27 (0.82)	6 (0.06)	9 (0.09)	59 (4.44)		16 (0.47)	8 (0.38)	47 (0.85)	21 (0.43)	58 (1.50)	24 (0.65)	6 (0.06)
Ostracoda					6 (0.06)		48 (1.28)					21 (0.28)			6 (0.12)
Copepoda	30 (0.40)	40 (1.00)		82 (10.64)	67 (6.06)	18 (0.73)	45 (1.62)		11 (0.16)		62 (3.23)	71 (4.29)	25 (0.58)	47 (1.29)	50 (1.44)
Isopoda		80 (1.80)		9 (0.09)	6 (0.06)										
Amphipoda		20 (0.20)				9 (0.09)						21 (0.43)	17 (0.17)	12 (0.12)	6 (0.06)
Insecta															
Plecoptera					33 (0.33)										
Ephemeroptera		60 (4.20)		45 (0.64)	39 (0.72)	45 (0.64)	34 (0.59)	24 (0.35)	26 (0.26)	8 (0.15)		14 (0.14)	17 (0.17)		44 (0.83)
Odonata															
Zygoptera							3 (0.03)				8 (0.08)				
Anisoptera							3 (0.03)	12 (0.12)			8 (0.08)	7 (0.07)		6 (0.06)	6 (0.06)
Trichoptera	30 (0.30)		10 (0.10)	18 (0.18)		55 (0.82)	38 (0.59)	18 (0.18)	58 (1.16)	15 (0.38)	15 (0.15)	43 (1.07)	42 (0.75)	35 (0.65)	17 (0.17)
Coleoptera					17 (0.22)				5 (0.05)		8 (0.08)	7 (0.07)	8 (0.08)	18 (0.18)	11 (0.22)
Diptera															
Chironomidae	100 (31.30)	100 (12.20)		82 (8.82)	100 (5.50)	82 (3.55)	100 (9.48)	71 (9.94)	74 (9.53)	77 (4.85)	92 (3.00)	86 (6.50)	92 (12.58)	94 (12.71)	100 (8.17)
Simuliidae	10 (0.10)	20 (0.20)			6 (0.11)		10 (0.66)	6 (0.06)	21 (0.79)			7 (0.07)	29 (0.07)	6 (0.67)	56 (0.06)
pupae		10 (0.10)			17 (0.22)			6 (0.06)			8 (0.08)			6 (0.06)	17 (0.17)
Other					6 (0.06)							7 (0.07)	17 (0.17)		
Miscellaneous	20 (0.20)	10 (0.10)					17 (0.24)					14 (0.14)			
Mean stomach fullness index	2.80	3.20	0.10	2.82	3.11	2.27	2.79	2.06	2.11	1.62	1.38	2.43	3.17	2.88	2.39

March 4 collection. At other dates both isopods and Anisoptera nymphs occurred infrequently, but comprised the bulk of the biomass in stomachs in which they were found. All other food item categories were rare.

Mean stomach fullness was considerably greater during winter and spring (December-May) than during summer and fall (June-November). The exception to this generalization was the unusual March 16 sample. In a total of 10 individuals only a single small caddisfly was found. In addition, the intestines of all 10 fish were void of food remains, suggesting that these fish had not fed the previous day. Since this date corresponds to the probable height of the reproductive period, it appears that gulf darters may not feed while actively spawning.

We have no data on the distribution or abundance of potential prey in Black Creek. However, the wide variety of prey categories consumed during most seasons suggests that *E. swaini* is an opportunistic feeder. Based upon what is known for other, similar darters, gulf darters probably rely almost exclusively on sight for feeding. This premise is supported by our visual observations of *E. swaini* during daylight hours.

Age and growth

Scale annuli were usually formed by the end of March, corresponding closely with the time of spawning (mid-February to late March). Scales consisted of zones of widely spaced ridges followed by zones of closely spaced ridges which correspond to periods of rapid (spring and summer) and slow (fall and winter) growth, respectively. The annulus was located at the outer edge of the closely spaced ridges. By using both methods (length frequency histograms and number of annuli) and by assuming that most fish were hatched during March, we were able to determine the actual age of fishes to within approximately one month. These analyses indicated that, at most, three year-classes were present in the drainage at any given time. During our study, 1975, 1976, 1977, and 1978 year-classes were represented in the population (Fig. 2). We found 51.1% of all *E. swaini* collected were less than 12 months old, 37.0% were 12–23 months old, 11.6% were 24–35 months old, and less than 0.5% were more than 35 months old.

Etheostoma swaini ranged in size from 9.9 to 50.5 mm SL. Fish $\leqslant$ 10 mm SL probably were not vulnerable to our sampling gear, since the 1978 year class did not occur in our samples until May, when we collected our smallest specimens (9.9 mm). Early growth was rapid and by the end of their first year fish had reached 61.8% of their eventual total length (1978 year class, mean SL at 12 months of age = 29.3 mm; Fig. 2). This first year growth rate is similar to that reported for *E. smithi* (31.9 mm SL; Page & Burr 1976), *E. barbouri* (29.2 mm SL; Flynn & Hoyt 1979), and *E. kennicotti* (29.3 mm SL; Page 1975). Growth rates diminished with age, decreasing dramatically during the second and third years. The mean standard lengths of 24-month-old and 36-month-old fishes taken during March were 37.9 and 41.9 mm SL, respectively.

Regression equations were derived for the length-weight relationship of males and females using logarithmic transformations. the length-weight relationship for males was: $\text{Log W} = -5.2008 + 3.2335 \text{Log L}$ ($n = 135$, $r = 0.97$), and for females: $\text{Log W} = -5.1860 + 3.2299 \text{Log L}$ ($n = 203$, $r = 0.97$).

Reproduction

Sexual dimorphism

Brilliant breeding coloration is common among darters (e.g. Winn 1958, Page & Burr 1976, Page 1975). Male and female *E. swaini* exhibit aqua-blue and orange coloration, which is most apparent prior to and during the breeding season (February-March). The coloration is especially striking in the male. Orange pigmentation is intense along the ventral surface from the pelvic fin to the anus and occurs in five distinct vertical bands on the posterior half of the lateral surface. Orange pigment is also found in a band along the base and middle portions of both dorsal fins, and in two spots on the anterior portion of the caudal fin. Aqua-blue pigmentation is found on the margins of both dorsal fins, as well as in band through the center of the first dorsal fin, between the orange bands. It also occurs on most of the pelvic fin, the anterior and

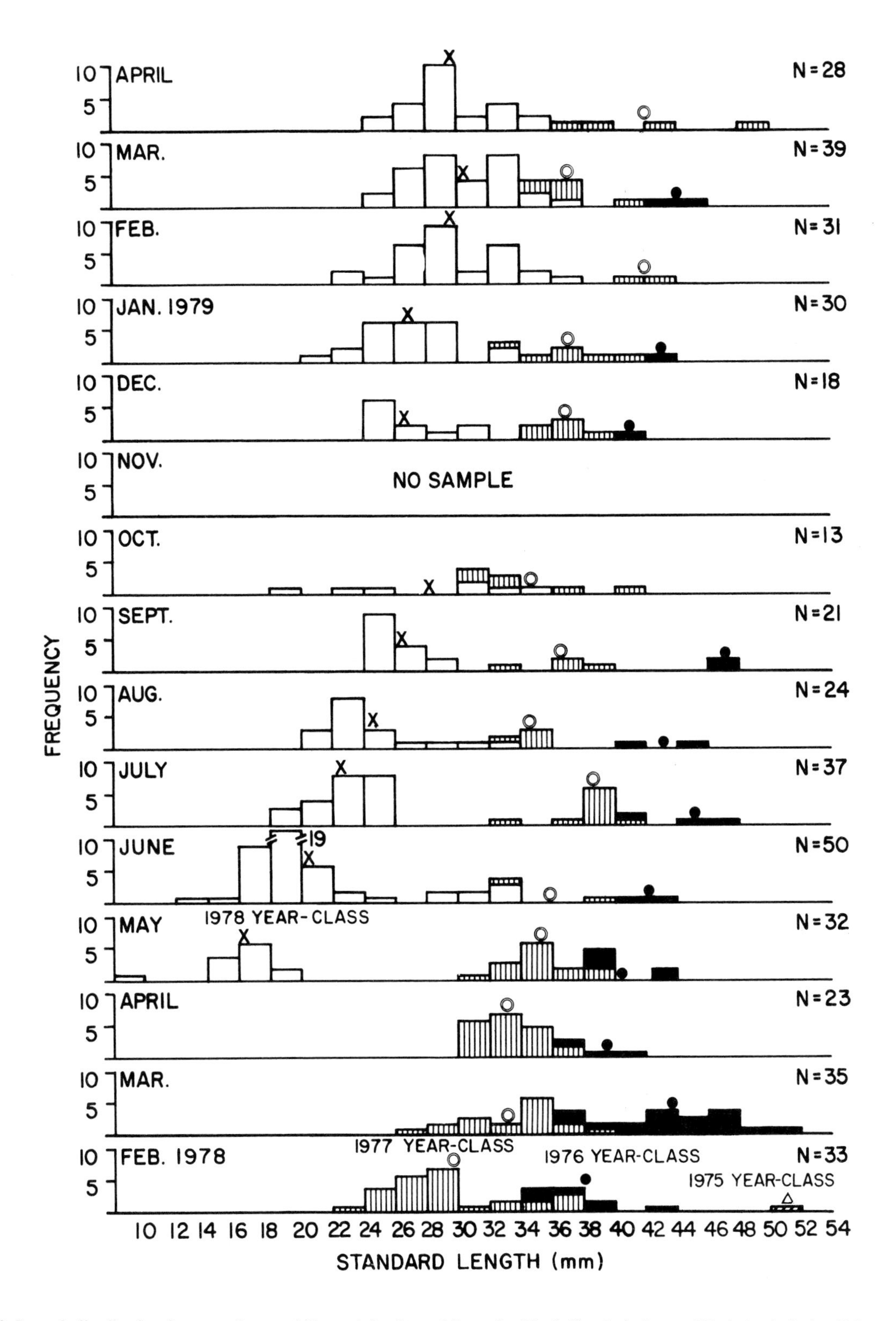

Fig. 2. Length distribution for year classes of *E. swaini* collected from the Black Creek drainage, Mississippi, during February 1978 to April 1979. Mean lengths for the 1978 year-class are represented by an X, 1977 year-class by an open circle, 1976 year-class by a closed circle, and the 1975 year-class by a triangle.

posterior edges of the anal fin membrane, and as a small patch of pigment on the upper and lower portions of the caudal fin base. Coloration varies somewhat between specimens.

The genital papilla is distended in female gulf darters prior to and during the breeding season.

Spawning behavior

On February 25, 1978 (1400 h) three female and two male *E. swaini* were collected from Perkins Creek and placed in a 75.68 *l* aquarium. The fish were apparently in spawning condition, the females swollen with eggs and the males brilliantly colored. The aquarium substrate consisted of separate sections of sand and small- to medium-size gravel. Several small rooted *Sparganium americanum* plants were also in the aquarium.

We observed spawning in the aquarium the following day from 1000 to 1200 h. The three females remained in one end of the aquarium, guarded closely by the somewhat smaller, but apparently dominant male. Two females began burrowing activity in the gravel substrate, pushing their nose into the gravel and submerging themselves with thrusts of the caudal fin. Their heads would quickly reappear, but the abdomen would remain buried. Males did not exhibit any burrowing behavior. After a female had burrowed, the dominant male would press himself against the top or side of the female and both would quiver rapidly, apparently spawning. The third female, although full of eggs, did not exhibit this burrowing behavior. She was, however, still guarded by the dominant male. This spawning behavior did not appear to be associated with the aquatic vegetation in the aquarium.

The dominant male was often distracted from attending the females to show aggressive behavior towards the other male. The dominant male repeatedly chased the submissive male to the far corner and quickly returned to the females. The dominant, more active male was much more darkly pigmented than the other male. His head and sides were a mottled dark brown to black color. Coloration in the second male would also darken when given a chance to follow a female. Similar courtship behavior has been reported by Winn (1958) for *E. flabellare*. The dominant male, although smaller in size, was very aggressive toward the other male. The dominant male fully displayed his dorsal fins when showing aggression and when courting the females, but often lowered them while at rest. The other male usually displayed a submissive posture, lowering his dorsal fins next to the body. The submissive male avoided the dominant male and was observed retreating behind the females. The females were guarded only during spawning. Neither the dominant male nor the females appeared to protect the area where spawning had taken place.

Spawning time and location

Time of spawning was determined from seasonal changes in gonadal somatic index (GSI) and field observations of ripe fish. Females with ripe and spent ovaries were taken during February, March, and April. Highest GSI values for both males and females occurred during March (Fig. 3). The spawning season extended from mid-February to late March at water temperatures from 5.5° C to 17.0° C (mostly between 7° C and 12° C). Considering the relatively short spawning season, the sharp decline in GSI in April, and the stage of early development of the immature eggs during March it

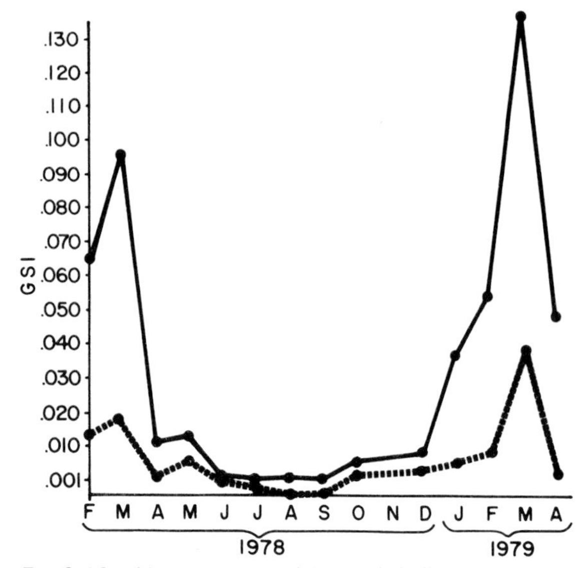

Fig. 3. Monthly average gonadal somatic indices for male and female *E. swaini* collected in the Black Creek drainage, Mississippi. Solid lines represent female GSI and broken lines represent male GSI.

appears that individual *E. swaini* spawn only once a season and resorb the immature eggs. This is further supported by the fact that young-of-the-year fishes began to occur in our collections only during one period of the year (Fig. 2). *Etheostoma swaini* appears to be the first percid to spawn each year in the Black Creek system.

Etheostoma swaini apparently spawns in gravel runs, although no eggs or spawning fish were observed in the field. On March 11, 1978 approximately 20 fish were taken from a swift, shallow (mean depth = 30 cm), gravel run in Black Tom Creek. This collection was the only one in which a large aggregation of males and females were taken together during the early spring 1978 samples. This was the only collection during the entire study in which this number of *E. swaini* were taken in such a confined area, and one completely lacking any aquatic vegetation or detrital material. Observations of spawning in the laboratory seem to substantiate the fact that spawning does occur in unvegetated, gravelly sites. Most of the males and females taken at this location were in prime spawning condition, with several of the females appearing to be already spent.

Fecundity and ova size

Numbers of mature ova (total fecundity) ranged from 7 to 90 in 14 fish examined (Table 2), with a mean of 39.0. Mean relative fecundity (number of mature ova per gram ovary-free body weight) was 43.0. Relative fecundity, although similar to total fecundity in this instance, is probably a more exact estimate of fecundity because it takes body weight into consideration. Similar numbers of mature ova have been reported for *E. kennicotti* (Page 1975), *E. smithi* (Page & Burr 1976), and *E. coosae* (O'Neil 1981).

Mature ova ranged from 1.1 to 2.0 mm in diameter, with a mean of 1.35 mm (n = 546). Mature ova contained considerable yolk material and a distinct oil globule.

Sex ratio

In the Black Creek drainage, females outnumbered males (Table 3) by a ratio of 59:41 ($\chi^2 = 11.94$; $p<0.005$). Sex ratios were based on all fish collected except those 1978 year-class fish whose sex could not be determined. Only in February 1978 and January 1979 were more males than

Table 2. Total and relative fecundity for 14 randomly selected *E. swaini* females taken from the Black Creek drainage, February and March 1978.

SL	Total fecundity[1]	Relative fecundity[2]
28.3	11	33.17
28.4	14	48.78
29.4	7	16.81
31.4	15	32.01
34.2	27	41.66
34.3	23	34.91
35.6	39	48.30
35.7	13	19.30
36.4	70	109.13
37.8	32	35.03
46.2	59	39.25
46.4	84	57.66
46.5	62	45.10
50.4	90	41.31
Mean	39.00	43.03
Standard Deviation	28.62	21.95

[1] Number of mature ova.
[2] Number of mature ova per gram, ovary-free body-weight.

Table 3. Sex ratio by month of *E. swaini* taken from the Black Creek drainage, Mississippi, February 1978–April 1979.

Month	Males	Females	Sex ratio
February	18	15	54:46
March	8	26	24:76
April	7	13	35:65
May	5	14	26:74
June	12	16	43:57
July	12	20	38:62
August	8	13	38:62
September	7	9	44:56
October	5	5	50:50
December	5	10	33:67
January	12	6	67:33
February	15	16	48:52
March	12	25	32:68
April	12	16	43:57
Total	139	203	41:59

females taken. Sex ratios by year-class were: 1978 year-class, 56:44; 1977 year-class, 58:42; 1976 year-class, 72:28; and 1975 year-class, 50:50. Sampling bias for one sex was probably not a major factor in the observed differences because females equaled or outnumbered males in 12 of the 14 months sampled, and we sampled all microhabitats at any particular sampling location.

Sex ratios in darters tend to vary greatly, but there appears to be a general tendency for males to exceed females (Lachner et al. 1950). Other *Etheostoma* species in which females outnumber males include *E. smithi* (Page & Burr 1976), *E. olmstedi* (Tsai 1972), *E. kennicotti* (Page 1975), *E. squamiceps* (Page 1974), *E. blennioides* (Wolfe et al. 1978), and *E. flabellare* (Lake 1936).

Acknowledgements

We would like to thank the Biology Department at the University of Southern Mississippi for the use of sampling equipment and facilities and to the patience and guidance of S.T. Ross. We also appreciate the assistance provided by Lucia O'Toole and Linda Paulson of the Gulf Coast Research Laboratory in typing the manuscript and figuring the graphs and histogram. Comments by two anonymous reviewers contributed greatly to the final text.

References cited

Baker, J.A. & S.T. Ross. 1981. Spatial and temporal resource utilization by southeastern cyprinids. Copeia 1981: 178–189.

Braasch, M. & P. Smith. 1967. The life history of the slough darter, *Etheostoma gracile* (Pisces: Percidae). Ill. Nat. Hist. Surv. Biol. Notes No. 58. 12 pp.

Burr, B. & L. Page. 1978. The life history of the cypress darter, *Etheostoma proeliare*, in Max Creek, Illinois. Ill. Nat. Hist. Surv. Biol. Notes No. 106. 15 pp.

Clark, K.E. 1978. Ecology and life history of the speckled madtom, *Noturus leptacanthus* (Ictaluridae). M.S. Thesis, Univ. Southern Mississippi, Hattiesburg. 128 pp.

Douglas, N.H. 1974. Freshwater fishes of Louisiana. Claitor's Publishing Division, Baton Rouge, Louisiana. 443 pp.

Flynn, R. & R. Hoyt. 1979. The life history of the teardrop darter, *Etheostoma barbouri* Kuehne and Small. Amer. Midl. Nat. 101: 127–141.

Howell, W.M. & R.D. Caldwell. 1965. *Etheostoma (Oligocephalus) nuchale*, a new darter from a limestone spring in Alabama. Tulane Stud. Zool. 12: 101–108.

Hubbs, C.L. & K.F. Lagler. 1964. Fishes of the Great Lakes region. Univ. Mich. Press, Ann Arbor. 213 pp.

Karr, J.R. 1964. Age, growth, fecundity, and food habits of fantail darters in Boone County, Iowa. Iowa Acad. Sci. Proc. 71: 274–280.

Lachner, E.A., E.F. Westlake & P.S. Handwerk. 1950. Studies on the biology of some percid fishes from western Pennsylvania. Amer. Midl. Nat. 43: 92–111.

Lake, C.T. 1936. The life history of the fantail darter, *Catonotus flabellaris flabellaris* (Rafinesque). Amer. Midl. Nat. 17: 816–830.

O'Neil, P.E. 1981. Life history of *Etheostoma coosae* (Pisces: Percidae) in Barbaree Creek, Alabama. Tulane Stud. Zool. Bot. 23: 75–83.

Page, L.M. 1974. The life history of the spottail darter, *Etheostoma squamiceps*, in Big Creek, Illinois and Ferguson Creek, Kentucky. Ill. Nat. Hist. Surv. Biol. Notes No. 89. 20 pp.

Page, L.M. 1975. The life history of the stripetail darter, *Etheostoma kennicotti*, in Big Creek, Illinois. Ill. Nat. Hist. Surv. Biol. Notes No. 93. 15 pp.

Page, L.M. 1980. The life histories of *Etheostoma olivaceum* and *Etheostoma striatulum*, two species of darters in central Tennessee. Ill. Nat. Hist. Surv. Biol. Notes No. 113. 14 pp.

Page, L.M. 1981. The genera and subgenera of darters (Percidae, Etheostomatini). Occas. Pap., Mus. Nat. Hist., Univ. Kansas 90: 1–69.

Page, L.M. & B.M. Burr. 1976. The life history of the slabrock darter, *Etheostoma smithi*, in Ferguson Creek, Kentucky. Ill. Nat. Hist. Surv. Biol. Notes No. 99. 12 pp.

Ramsey J.S. & R.D. Suttkus. 1965. *Etheostoma ditrema*, a new darter of the subgenus *Oligocephalus* (Percidae) from springs of the Alabama River Basin in Alabama and Georgia. Tul. Stud. Zool. 12: 65–77.

Starnes, W.C. 1980. *Etheostoma swaini* (Jordan), gulf darter. pp. 699. *In*: D.S. Lee et al. (ed.) Atlas of North American Freshwater Fishes, N.C. State Mus. Nat. Hist., Raleigh.

Swift, C.C., R.W. Yerger & P.R. Parrish. 1977. Distribution and natural history of the fresh and brackish water fishes of the Ochlockonee River, Florida and Georgia. Bull. Tall Timbers Res. Sta. No. 20. 111 pp.

Tsai, C. 1972. Life history of the eastern johnny darter, *Etheostoma olmstedi* Storer, in cold tailwater and sewage-polluted water. Trans. Amer. Fish. Soc. 101: 80–88.

Ultsch, G.R., H. Boschung & M.J. Ross. 1978. Metabolism, critical oxygen tension, and habitat selection in darters (*Etheostoma*). Ecology 59: 99–107.

Winn, H.E. 1958. Observations on the reproductive habits of darters (Pisces-Percidae). Amer. Midl. Nat. 59: 190–212.

Wolfe, G.W., B.H. Bauer & B.A. Branson. 1978. Age and growth, length-weight relationships, and condition factors of the greenside darter from Silver Creek, Kentucky. Trans. Ky. Acad. Sci. 39: 131–134.

Originally published in Env. Biol. Fish. 11: 121–130

Habitat partitioning among five species of darters (Percidae: *Etheostoma*)

Matthew M. White[1] & Nevin Aspinwall[2]
Department of Biology, St. Louis University, St. Louis, MO 63103, U.S.A.

Keywords: Ecology, Fish community, Perciformes, Resource partitioning

Synopsis

An analysis of habitat partitioning among five species of darters *(Etheostoma blennioides, E. caeruleum, E. spectabile, E. tetrazonum* and *E. zonale)* suggests that differential use of vegetation types is of primary importance in the microhabitat separation of these species. Substrate size, depth and current appear to separate only one species consistently. A comparison of data from localities where all species were collected with localities where a one of the five species was absent suggests some degree of habitat displacement.

Introduction

Despite early studies which concluded that trophic separation was more important in aquatic communities (Schoener 1974), recent work suggests that spatial dimensions play a more important role in the structuring of fish communities (Mendelson 1975, Werner et al. 1977). There appears to be a correlation between stream habitat complexity and fish species diversity with small stream fishes being habitat specialists (Gorman & Karr 1978).

Studies on cyprinid communities (Mendelson 1975, Baker & Ross 1981) and two species of darters (Smart & Gee 1979) have shown that vertical separation (position in the water column) was important in the ecological separation of members of these communities. Differential association with vegetation has been shown to be important as well (Baker & Ross 1981). Competition for habitat space has been implicated as a cause of the allopatric distributions of a group of slab-pool darters (Page & Schemske 1978).

The Meramec River drainage of the Ozark Upland region of Missouri supports a diverse darter fauna (12 species, Pflieger 1975). This study was initiated to determine how a group of these species partition certain aspects of the stream environment. Five species of darters of the genus *Etheostoma,* all common throughout most of the drainage, were chosen for study: *E. blennioides,* greenside darter, *E. caeruleum,* rainbow darter, *E. spectabile,* orangethroat darter, *E. tetrazonum,* Missouri saddled darter, *E. zonale,* banded darter. (For general habitat descriptions see Fahy 1954, Winn 1958, Pflieger 1975.) We examined the importance of substrate size, depth, current velocity and vegetation type in the ecological separation of these species.

[1] Present address: Department of Biology, Virginia Polytechnic Institute and State University, Blacksburg, VA 24061, U.S.A.
[2] To whom reprint requests should be sent.

David G. Lindquist & Lawrence M. Page (ed.), Environmental biology of darters. ISBN 90-6193-506-7

Materials and methods

Streams in the Ozark Upland region are usually clear with moderate to high gradient and are characterized by alternating pools and riffles. Fifteen localities were selected to reflect the varying degrees of stream size and drainage location. The first five localities listed are those where all five species were collected. Numbers in parentheses refer to the number of samples taken at each locality: 1. Meramec River at Gray Summitt, Franklin County, 4 September 1977 (17); 2. Huzzah Creek at Rte. E, Crawford County, 15 June 1978 (9); 3. Bourbeuse River at Ryker Ford, Franklin County, 6 August 1978 (19); 4. Big River at Rte. B, Jefferson County, 26 August 1978 (20); 5. Meramec River at Times Beach, St. Louis County, 21 September 1978, 7 October 1978 (47); 6. Meramec River at Rte. K, Franklin County, 9 June 1978 (6); 7. Courtois Creek at Bass Canoe Rentals, Crawford County, 14 June 1978 (6); 8. Courtois Creek at Rte. 8, Crawford County, 14 June 1978 (6); 9. Big River at Rte. W, Franklin County, 17 June 1978 (6); 10. Little Meramec River at Rte. N, Crawford County, 19 June 1978 (12); 11. Huzzah Creek at St. Louis University Biological Station, Crawford County, 5 July 1978 (11); 12. Meramec River at Rte. 8, Crawford County, 6 July 1978 (7); 13. Meramec River at Hwy. 19, Crawford County, 6 July 1978 (9); 14. Meramec River at Rte. N, Franklin County, 2 September 1978 (7); 15. LaBarque Creek at Rte. FF, Jefferson County, 30 September 1978 (16).

Sampling occurred in September 1977 and from June to November, 1978. Field work was not attempted during the intervening period due to seasonal flooding and to avoid sampling the reproductive habitat. Fishes were collected with a 1.5 m minnow seine with 3 mm mesh. The number of samples taken at any locality varied with the size of the stream but were taken within a 100 m stretch of stream. Samples were taken at random based on a grid. Each locality was sampled on a single date with the exception of a large downstream locality which required two different sampling times. Where seining was not possible, fishes were visually censused by snorkeling.

For each 1.5 m sample, the habitat was characterized with respect to four variables: substrate size, depth, current velocity and vegetation type. Surface current velocity was measured by timing a floating chip. Four classes of vegetation reflected the various types of aquatic vegetation common to stream communities: filamentous algae (primarily *Cladophora)*, emergent vegetation (*Justicia americana*), submergent vegetation (primarily *Potamogeton*) and 'no' vegetation.

The principal component analysis was performed using the NTSYS computer package (V.P.I. & S.U.). A weighted mean (weighted for density of each species in each sample) of substrate size, depth and current velocity as well as occurrence (frequency of samples where a species was found in a particular vegetation class) and density (fish per square meter) in the vegetation classes was computed for each species from data at all fifteen localities and five localities where all five species occured. The values were standardized ($Y = 0$, $s = 1.0$) before use in the principal component analysis. Niche overlap values (Ojk) were computed using the index developed by Pianka (1973). Multidimensional overlap was determined using the multiplication procedure.

Results

Observations were made on a total of 198 samples from the fifteen localities. *Etheostoma caeruleum* was the most abundant species throughout the drainage (83 samples, 5.1 individuals per sample), but at the five localities where all species occurred, *E. tetrazonum* dominated (50 samples, 4.3 individuals per sample).

The mean values for the habitat characteristics for each species appear in Table 1. A comparison of multidimensional niche overlap values indicated that *E. spectabile* exhibited the smallest overlap with the other species (0.085 to 0.165). The other four species had relatively high overlap values (0.824 to 0.841) except for *E. caeruleum* with *E. tetrazonum* and *E. zonale* (0.702 and 0.667, respectively).

Principal component analysis separated *E. cae-*

Table 1. Mean character values for the five species of darters from all 15 localities.

Character	*E. blennioides*	*E. caeruleum*	*E. spectabile*	*E. tetrazonum*	*E. zonale*
Substrate (cm)	11.20	10.03	3.89	10.45	11.05
Depth (cm)	28.80	33.80	19.12	24.83	26.96
Current (m s^{-1})	0.64	0.60	0.26	0.67	0.61
Veg1-occurrence	0.46	0.39	0.22	0.52	0.39
-density	3.37	3.06	3.33	4.05	2.36
Veg2-occurrence	0.14	0.20	0.09	0.19	0.08
-density	3.33	9.76	4.40	4.46	2.25
Veg3-occurrence	0.19	0.06	0.03	0.17	0.33
-density	5.00	1.80	1.00	3.23	2.26
Veg4-occurrence	0.28	0.38	0.64	0.26	0.27
-density	2.46	5.84	8.38	2.25	1.83

ruleum and *E. spectabile* from the other species (Fig. 1). Heaviest loadings on the first component were substrate size, current, occurence in filamentous algae, and occurrence and density in vegetation free areas (Table 2). Occurrence and density in emergent vegetation loaded heavily on the second component, and density in filamentous algae loaded heavily on the third component. The cutoff value used to determine relative importance was 0.700. *Etheostoma blennioides, E. tetrazonum* and *E. zonale* grouped closely (Fig. 1).

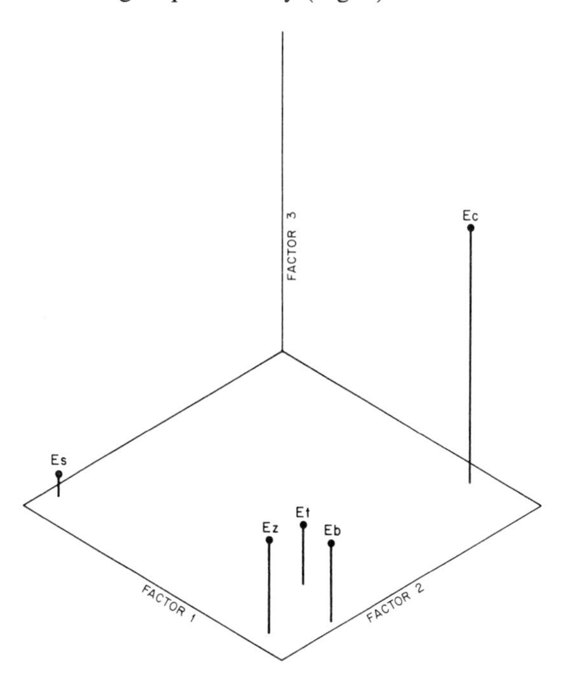

Fig. 1. Plot of the principal component analysis for the mean data from all 15 localities. Abbreviations: *Eb E. blennioides, Ec E. caeruleum, Es E. spectabile, Et E. tetrazonum, Ez E. zonale.*

In order to address the importance of the habitat characteristics in areas where all five species were present, only data from localities 1-5 were analysed (Table 3). Separation is greater than that seen in the analysis of all fifteen localities, especially concerning *E. blennioides, E. tetrazonum* and *E. zonale* (Fig. 2). Component two contributed heavily to their separation, especially the occurence and density in emergent vegetation and occurence in submergent vegetation. To further analyse habitat similarities of these three species, the relationships between occurrence and density in filamentous algae and submergent vegetation were examined (Fig. 3). Figures 3a and 3c represent areas of complete sympatry; 3b and 3d represent localities where *E. zonale* was extremely rare or absent. In the absence of *E. zonale* there were relatively minor changes in the occupancy of fila-

Table 2. Loadings of the mean habitat characteristics for all fifteen localities on principal components (PC) 1 to 3.

	PC1	PC 2	PC 3
Substrate size	0.971	−0.025	0.201
Depth	0.612	−0.484	0.589
Current	0.982	−0.154	0.073
Veg1-occurrence	0.938	−0.198	−0.260
-density	0.032	−0.502	−0.860
Veg2-occurrence	0.406	−0.898	−0.123
-density	−0.155	−0.907	0.388
Veg3-occurrence	0.683	0.697	0.159
-density	0.764	0.059	−0.402
Veg4-occurrence	−0.992	−0.067	−0.035
-density	−0.933	−0.331	0.089
% of variation	57.0	25.3	14.2

Table 3. Loadings of the mean habitat characteristics from the five localities of complete sympatry on principal components (PC) 1 to 3.

	PC 1	PC 2	PC 3
Substrate size	0.938	–0.157	0.207
Depth	0.950	–0.209	0.227
Current	0.647	–0.430	–0.629
Veg1-occurrence	0.903	0.278	–0.230
-density	0.304	0.601	–0.688
Veg2-occurrence	0.471	0.871	–0.098
-density	0.340	0.801	0.370
Veg3-occurrence	0.210	–0.924	0.227
-density	0.595	–0.681	0.185
Veg4-occurrence	–0.994	0.069	0.089
-density	0.339	0.520	0.780
% of variation	44.9	33.5	16.9

mentous algae (Fig. 3a, 3b). In contrast, *E. blennioides* and *E. tetrazonum* exhibit a marked increase in the occupancy of submergent vegetation in the absence of *E. zonale* (Fig. 3c, 3d).

Discussion

Although Gorman & Karr (1978) suggested that substrate size, depth and current are important in the microhabitat specialization of stream fishes, our results suggest that these play a lesser role than does vegetation in the habitat separation of the darters studied. However, *E. spectabile* did consistently separate on the basis of substrate size and current velocity. Depth did not appear to be important despite its role in other fish communities (Smart & Gee 1979, Baker & Ross 1981).

Differential occupation of vegetation was most important in the habitat separation of the remaining four species. Since *Cladophora* is a common characteristic of riffles in the Meramec River system, it is not surprising that all species were associated with it to some extent (Table 1). *Etheostoma spectabile* was frequently found, often abundantly, on riffles with no vegetation. Often these were small streams with intermittent flow. The predominat species in emergent vegetation were *E. caeruleum* and *E. tetrazonum* with the former exhibiting the highest density of any species.

The habitat characteristics of *E. blennioides, E. tetrazonum* and *E. zonale* were similar (multidimensional overlap values from 0.824 to 0.841). Pflieger (1975) concluded that *E. blennioides* and *E. zonale* were the most closely associated of any Missouri darters and that there appeared to be no significant difference in their habitats. However, an important difference among these species appears to be in the occupation of submergent vegetation. Where it was common, *E. zonale* is the species associated with *Potamogeton* (Fig. 3c). At localities where *E. zonale* is absent or very rare, *E. blennioides* and *E. tetrazonum* show a marked increase in occurence and density in *Potamogeton*. Though not conclusive, this suggests some degree of competitive displacement on the part of *E. zonale*.

Changes in abundance with stream size and drainage position may also be important in the separation of the five species. Four of the five localities of complete sympatry were the largest and most downstreams. Here, *E. tetrazonum* and *E. zonale* were the most abundant species. Pflieger (1975) noted that *E. tetrazonum* appears to avoid headwater and small streams, preferring larger streams with continuous strong flow. *Etheostoma blennioi-*

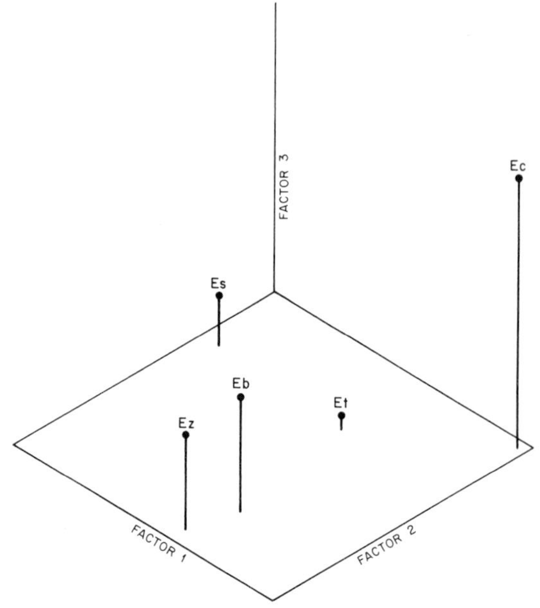

Fig. 2. Plot of the principal component analysis for mean data from the five localities where all species were collected. Abbreviations as in Figure 1.

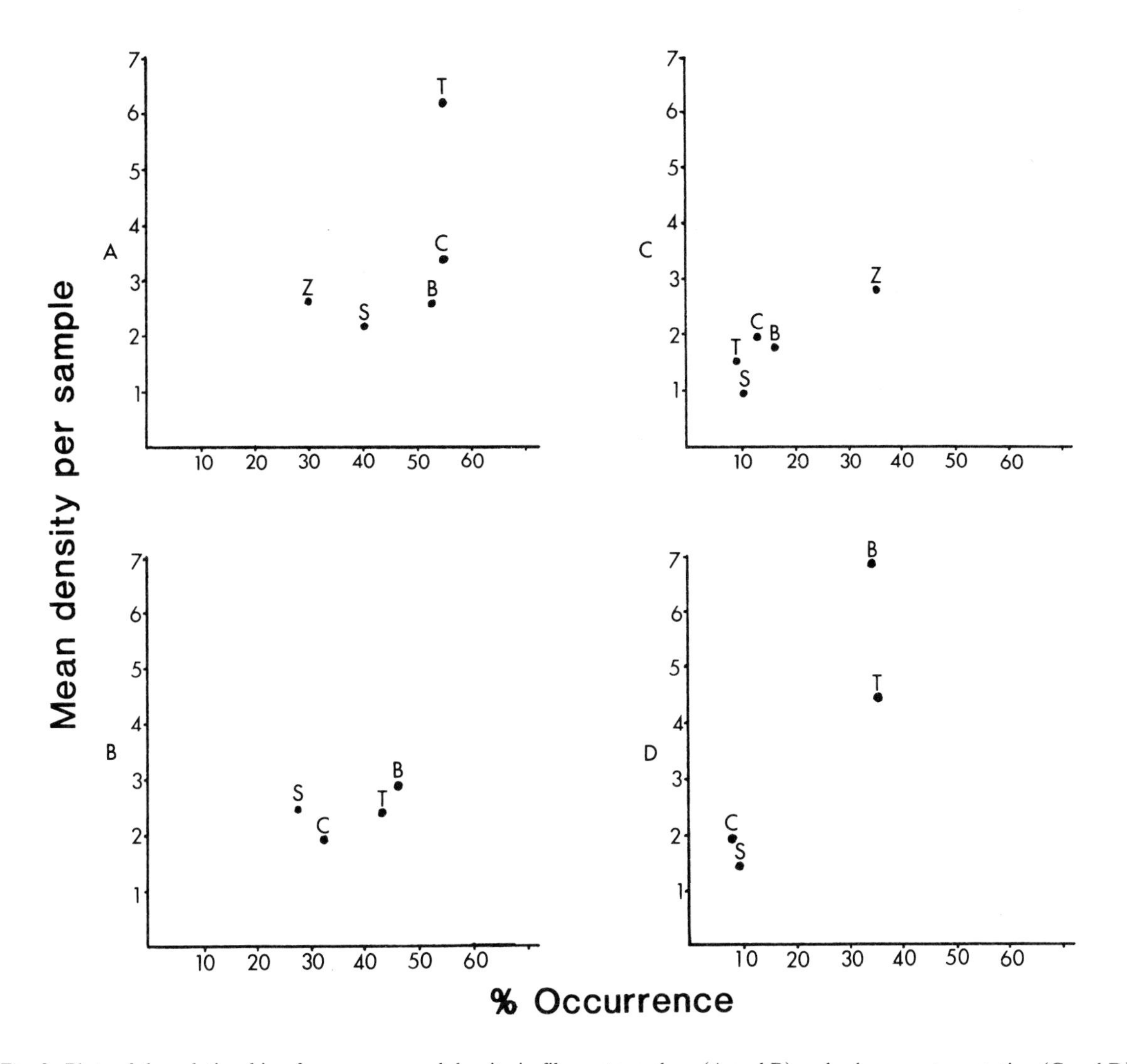

Fig. 3. Plots of the relationship of occurrence and density in filamentous algae (A and B) and submergent vegetation (C and D) in habitats with (A and C) and without *E. zonale* (B and D). Abbreviations as in Figure 1.

des and *E. caeruleum* were common throughout the drainage.

Although little is known of the winter habits of darters, sampling during other times of the year, when vegetation has died back, may reveal quite different partitioning patterns. Differences noted in vegetation use may be related to differential foraging behaviors or efficiencies, but this cannot be addressed at this time. Clearly an analysis of the seasonal aspects of habitat selection, prey availability, diet and foraging behavior is necessary to address the relative importance of spatial, trophic and temporal relationships among the darters. This study does suggest the importance, at least seasonally, of vegetation in the structuring of darter communities.

Acknowledgements

Field assistance was provided by J. Foley. Our thanks to K. Hilu and T.A. Grudzien for help in data analysis. J. Cranford and R. Andrews offered constructive criticisms on the manuscript. Thanks to D.G. Lindquist for allowing us to present this work at the darter symposium. Special thanks go to J.W. Fournie for field assistance and endless moral support during the length of this study.

References cited

Baker, J.A. & S.T. Ross. 1981. Spatial and temporal resource utilization by southeastern cyprinids. Copeia 1981: 178–189.

Fahy, W.E. 1954. The life history of the northern greenside darter, *Etheostoma blennioides blennioides* Rafinesque. J. Elisha Mitchell Sci. Soc. 70: 139–205.

Gorman, O.T. & J.R. Karr. 1978. Habitat structure and stream fish communities. Ecology 59: 507–515.

Mendelson, J. 1975. Feeding relationships among species of *Netropis* (Pisces: Cyprinidae) in a Wisconsin stream. Ecol. Monogr. 45: 199–230.

Page, L.M. & D.W. Schemske. 1978. The effect of interspecific competition on the distribution and size of darters of the subgenus Catonotus (Percidae: *Etheostoma).* Copeia 1978: 406–412.

Pflieger, W.L. 1975. The fishes of Missouri. Missouri Depart. of Cons. Jefferson City. 343 pp.

Schoener, T.W. 1974. Resource partitioning in ecological communities. Science 185: 27–39.

Smart, H.J. & J.H. Gee. 1979. Coexistence and resource partitioning in two species of darters (Percidae), *Etheostoma nigrum* and *Percina maculata.* Can. J. Zool. 57: 2061–2071.

Werner, E.E., D.J. Hall, D.R. Laughlin, D.J. Wagner, L.A. Wilsman E F.C. Funk. 1977. Habitat partitioning in a freshwater fish community. J. Fish. Res. Board Can. 34: 360–370.

Winn, H.E. 1958. Comparative reproductive behavior and ecology of fourteen species of darters (Pisces: Percidae). Ecol. Monogr. 28: 155–191.

Received 23.8.1982 *Accepted 26.7.1983*

Life history of the naked sand darter, *Ammocrypta beani*, in southeastern Mississippi

David C. Heins[1] & J. Russell Rooks
Department of Biology, Millsaps College, Jackson, MS 39210, U.S.A.

Keywords: Age, Breeding period, Egg size, Fecundity, Fish, Growth, Maturity, Ovarian weight, Percids, Sex ratio

Synopsis

The reproduction of *Ammocrypta beani* in the Jourdan River drainage, Hancock County, Mississippi, was studied from 1970–1976. In 1971, reproduction occurred from late March into early October. Data from other years support these observations. Females from Hickory Creek produced more mature ova than females from Catahoula Creek. These differences may be due to seasonal, annual, habitat, or genetic differences. Counts of mature ova for both localities ranged from 27–120 for females 35–48 mm SL. Mature ova did not differ in size between the two localities. The mean mature ovum diameter was 0.92 mm. Maximum lifespan is about 2 years. Maturity occurs at about one year.

Introduction

Information on the natural history of the naked sand darter, *Ammocrypta beani* Jordan, is almost lacking. The distribution, habitat, and common species associates of *A. beani* were described by Williams (1975), who also noted the dates of collections containing gravid females and the number and size of mature ova present in them. Our study is the first detailed consideration of the reproduction, age and growth of *A. beani*.

Ammocrypta beani is a member of the *beani* species group in the subgenus *Ammocrypta* (Bailey & Gosline 1955, Williams 1975, Page 1981). It occurs in streams below the fall line in the Mobile Basin and westward along the Gulf Coastal Plain through the Lake Pontchartrain drainages to the eastern tributaries of the Mississippi River in southern Mississippi (Williams 1975). It is also found in the Hatchie River drainage, a Mississippi River tributary in southwestern Tennessee (Williams 1975). *Ammocrypta beani* inhabits shifting sand in clean, clear streams varying in size from creeks 3.0–4.6 m wide to rivers with moderate flow; however, it is more common in large streams in water 0.2–1.2 m deep (Williams 1975).

Materials and methods

Monthly collections of *A. beani* were taken from Catahoula Creek below the mouth of Dead Tiger Creek, Jourdan River system, Hancock County, Mississippi, from February through December 1971. Supplementary collections were taken from Catahoula Creek and Hickory Creek (another tributary in the drainage) at State Hwy 43 in 1970, 1972, and 1974–76. Collections were made with 3 m long, 4.8 mm and 3.2 mm Ace-mesh seines. Specimens

[1] Present address: Department of Biology, Tulane University, New Orleans, LA 70118, U.S.A.

David G. Lindquist & Lawrence M. Page (ed.), Environmental biology of darters. ISBN 90-6193-506-7

were preserved in 10% formalin and stored in 43% isopropyl alcohol. They are deposited in the Mississippi State University Collection of Fishes (1970–72) and the Tulane University Museum of Natural History (1974–76). Climatic records for the nearest recording station (Picayune, MS) were obtained from the National Climatic Center, Asheville, North Carolina.

The reproductive condition of females was determined by measurements of ova and ovarian weights and gross examination of ovaries and ova. Ovaries were removed from each of 10 adult females (when available) chosen at random from the 1971 collections. The diameter of one of the largest ova was measured. Because preserved ova were not spherical, diameters were estimated by averaging measurements of the largest and smallest dimensions. Measurements were made to the nearest 0.05 mm using an ocular micrometer. Specimens and ovaries were dried to constant weight at 100–105° C and weighed to the nearest 0.001 g; ovarian weight was calculated as a percentage of somatic body weight. Specimens were eviscerated before drying to eliminate a possible source of variation in the data.

Gross assessments of the reproductive condition of females were based on the classification of females into one of the following stages of ovarian condition: (1) Latent (LA)-ovaries small, thin, transparent to slightly translucent. Larger, developing ova, if present, yolkless or with nuclei visible; (2) Early maturing (EM)-ovaries small to moderate in size, translucent to white in color. Larger eggs relatively small with nuclei obscured by yolk deposition, often numerous, and becoming white in color; (3) Late maturing (LM)-ovaries moderate in size to greatly enlarged, filling a large portion of the body cavity, white to cream color. Larger maturing ova often as large as mature ova but not easily differentiated from smaller maturing ova, white or cream to sometimes yellow; (4) Mature (MA)-ovaries greatly enlarged, filling a major portion of the body cavity and frequently distending the abdomen. Mature or ripe ova present, easily differentiated from maturing ova on the basis of size and color. Mature ova opaque, cream to yellow in color, becoming translucent with small oil globules visible. Ripe ova transparent yellow or amber with large oil globules, often with vitelline membranes elevated; (5) Partially spent (PS)-ovaries noticeably smaller than mature ovaries, relatively small number of mature ova present.

Analysis of reproductive condition in males was based primarily on gross examination of the testes, but tuberculation was considered. Reproductively mature males had enlarged, opaque white testes. Latent males had small, transparent to translucent testes.

All individuals capable of producing gametes were considered sexually mature adults. More specifically, females were considered sexually mature if in one ovary they had at least three enlarged eggs with the nucleus obscured (or nearly so) by yolk deposition. Females classified as MA or PS were considered reproductive.

Fecundity was determined by direct counts of mature ova in ovaries of reproductively mature females. Only specimens with their body cavities filled by the ovaries (often distending the abdomen) were examined. Ten mature ova from each female were chosen at random and their diameters measured to determine mean ovum diameter. Eggs were not measured for females which had ova with elevated vitelline membranes.

Estimates of age were based on length frequency analyses. All individuals in each collection analyzed were measured to the nearest 0.1 mm standard length (SL). Length frequency histograms were prepared by plotting the percentage frequency for each 1-mm size group.

Results

Study areas and species associates

Heins & Clemmer (1975) gave a general description of Catahoula Creek and the climate of the geographical area. They also provided some data on the ecological conditions in their primary study area (Station 7, Catahoula Creek) where the collections of *A. beani* examined for this study were taken. Hickory Creek is similar to Catahoula Creek.

In the sampling area, Catahoula Creek was normally about 12–30 m wide and was usually less than about 1.0–1.2 m deep. The stream typically had swift, shallow runs alternating with flowing reaches with moderate current. However, Catahoula Creek was subject to rapid increases in discharge and flow rates which altered these conditions following heavy rains. The substrate was composed of sand with varying amounts of gravel admixed in isolated areas. Stream gradient was 0.59 m km^{-1}. Water quality measurements made in the sampling area are summarized by Heins & Clemmer (1975). The riparian region was composed of wooded banks and large, open sandbars that occupied about 50% of the total shoreline.

Ammocrypta beani was the third most abundant species (1326 specimens) taken in the study area between February 1971 and March 1972 (Heins & Clemmer 1975). It was commonly taken with *Notropis longirostris,* the most abundant species (5513 specimens) collected in the area. *Notropis venustus* was the second most abundant species (4599 specimens). However, *A. beani* was most closely associated with *N. longirostris* as both inhabit the same open sand bottoms (Heins & Clemmer 1975). Bailey et al. (1954) also noted a common association between *A. beani* and *N. longirostris*. However, the fish they studied has been referred to *Ammocrypta bifascia* (Williams 1975). The second most abundant darter was *Percina nigrofasciata* (544 specimens). Other darters were taken infrequently in small numbers (number of specimens in parentheses): *Etheostoma stigmaeum* (13), *Etheostoma histrio* (3), *Etheostoma swaini* (2), and *Etheostoma fusiforme* (1). A total of 34 other species was taken with *A. beani* (Heins & Clemmer 1975).

Reproductive periodicity in females

In 1971, reproduction occurred from late March into early October. Prior to the start of the reproductive season in February, diameters of larger ova and weights of ovaries averaged 0.36–0.39 mm and 1.7–2.3% somatic weight, respectively (Fig. 1). In early February, 85% (n = 47) of the adult (≥33 mm SL) females were EM; 15% were LM. By late February, 43% of the females (n = 7) were LM; the rest were EM. The first reproductive females were collected in late March when 70% (n = 10) were MA or PS; the others were LM. Reproductive females were taken from late March until early October. During these months, diameters of larger ova and weights of ovaries averaged 0.43–0.98 mm and 1.7–8.6% somatic weight, respectively. At the first of October, 13% (n = 47) of the females were reproductive (MA or PS); the rest were non-reproductive (LA or EM), indicating that reproduction had nearly ended. However, most adult females were reproductive throughout the breeding season. For example, from April through August 90% of the adult females (n = 240) were MA or PS; 10% were EM or LM. Water temperatures recorded with the collections taken during the reproductive season were 17–29° C. After the breeding season in November, all females (n = 81) were LA or EM. In November and December, diameters of larger ova and weights of ovaries averaged 0.12–0.30 mm and 1.1–1.3% somatic weight, respectively.

Data from other years (Table 1) support these observations. In 1970, reproduction had nearly ended by late September and was completed before mid October. In 1972, reproduction had begun before late March. In 1974, reproduction had begun before our earliest collection in mid March and continued into mid September, presumably ending sometime in late September. Reproduction began in late February of 1975 with the percentage of reproductive females increasing until late March when most adult females were reproductive and continued into late September, presumably ending in early October. Heins (1979) found that in 1975, *Notropis longirostris* from the Jourdan River drainage began reproducing during February. This was a relatively early initiation and was associated with relatively high mean monthly air temperatures during the first part of the year. In 1976, reproduction in *A. beani* began before early April. Data on the reproductive condition of males is in agreement with the data described above. The data on the start of the 1975 reproductive season illustrate a typical maturation pattern with a progressively greater percentage of individuals becoming re-

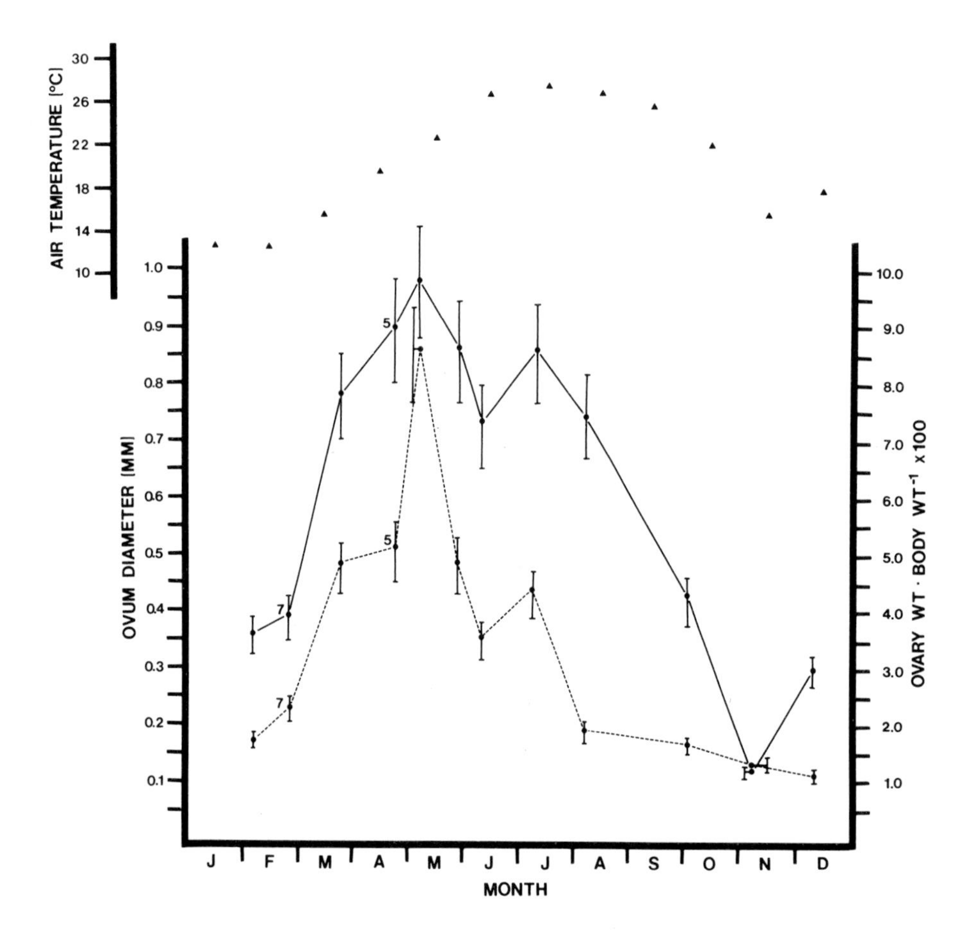

Fig. 1. Mean diameter of larger ova (solid line) and mean ovarian weight (% body weight; dashed line) for *Ammocrypta beani* from Catahoula Creek, Hancock County, Mississippi, in 1971. Vertical lines represent ±1 S.E. mean; n = 10 except where indicated. Mean monthly air temperature in 1971 (triangles; Picayune, Mississippi) is also shown.

productive. Similarly, the data for 1974–75 illustrate typical changes in reproductive condition for the populations at the end of the reproductive season. Williams (1975) reported gravid females between early June and late August.

Mature and partially spent females contained two complements of vitellogenic ova: large, mature ova and smaller, maturing ova. A bimodal distribution of yolk-bearing ova is considered indicative of a variable number of spawnings with complements of ova released periodically over an extended reproductive season (Hickling & Rutenberg 1936, Prabhu 1956, Qasim & Qayyum 1961, MacGregor 1970).

Reproductive periodicity of males

In 1971 the reproductive cycle in males coincided with the cycle in females. By late February, most adult males were reproductively mature (77%, n = 13) and tuberculate. Adult males (n = 18) in late March were mature. Almost all were tuberculate. Adult males remained reproductively mature and most were highly tuberculate from March through early August (n = 270). By early October, a large portion of the adult males were latent (36%, n = 81). Mature males were still tuberculate. In early November 53% (n = 88) of the males were latent; many of the mature males were close to being classified latent. All of them were weakly tuberculate or had no tubercules.

Table 1. Gross assessments of the reproductive condition of adult (≥33 mm SL) *Ammocrypta beani* females from Catahoula Creek and Hickory Creek, Jourdan River drainage, Hancock County, Mississippi, during various years. The number of females examined (N) and the percentage of individuals in each stage of reproductive condition are given: LA = latent, EM = early maturing, LM = late maturing, MA = mature, PS = partially spent.

Year	Date(s)	N	LA	EM	LM	MA-PS
1970	19 September	12	41.7	41.7	0.0	16.7
	18 October	12	75.0	25.0	0.0	0.0
1972	18 February	13	7.7	76.9	15.4	0.0
	25 March	8	0.0	12.5	12.5	75.0
1974	14 March	39	0.0	10.3	2.6	87.2
	3 September	29	10.3	10.3	6.9	72.4
	16 September	17	5.9	64.7	0.0	29.4
	26 September	24	16.7	75.0	4.2	4.2
	15 October	30	26.7	70.0	3.3	0.0
1975	18 February	14	0.0	50.0	50.0	0.0
	28 February	12	8.3	41.7	33.3	16.7
	7 March	18	0.0	33.3	27.8	38.9
	10 March	25	0.0	16.0	40.0	44.0
	22 March	13	0.0	7.7	38.5	53.8
	29 March	19	0.0	10.5	0.0	89.5
	1 September	21	0.0	0.0	0.0	100.0
	28 September	23	21.7	26.1	8.7	43.5
	4 November	8	25.0	75.0	0.0	0.0
1976	11 February	17	0.0	100.0	0.0	0.0
	5 March	5	0.0	0.0	100.0	0.0
	6 April	24	0.0	0.0	0.0	100.0

Fecundity

There was a highly significant correlation ($r = 0.779$, $n = 32$, $p<0.01$) between number of mature ova and SL among females taken from Catahoula Creek in 1971. Females 35–46 mm SL contained 27–101 ova. The linear equation expressing the relationship between fecundity (F) and SL (Fig. 2) is $F = 144.2256 + 5.1632(SL)$. Our samples from Catahoula Creek included few large, mature females. Therefore we also examined females taken from Hickory Creek during 1976. The correlation between number of mature ova and SL was highly significant ($r = 0.746$, $n = 22$, $p<0.01$). Counts of mature ova were 37–120 for females 35–48 mm SL. The linear relationship between fecundity and SL is $F = -89.8417 + 4.0313$ (SL). Covariance analyses (Snedecor & Cochran 1967) of the linear regression lines expressing the relationship between counts of mature ova and body size for females from the two localities did not differ significantly in slope but did differ significantly in elevation (Table 2). Williams (1975) observed 52–71 mature eggs in females 39–50 mm SL taken in June and early July. Williams (1975) also noted a reduction in the number of mature ova during the reproductive season of *A. beani*. Our gross observations of ovarian condition are in agreement with his. Specimens examined from Catahoula Creek were taken between early and late May while those from Hickory Creek were taken in April and early May. The differences in number of mature ova that we have observed may result from a seasonal reduction in the number of eggs. However, it may also be due to annual, habitat, or possibly genetic differences.

Log-log transformations yielded r^2 values that were similar to those for our linear equations, indicating that the relationships between ovum num-

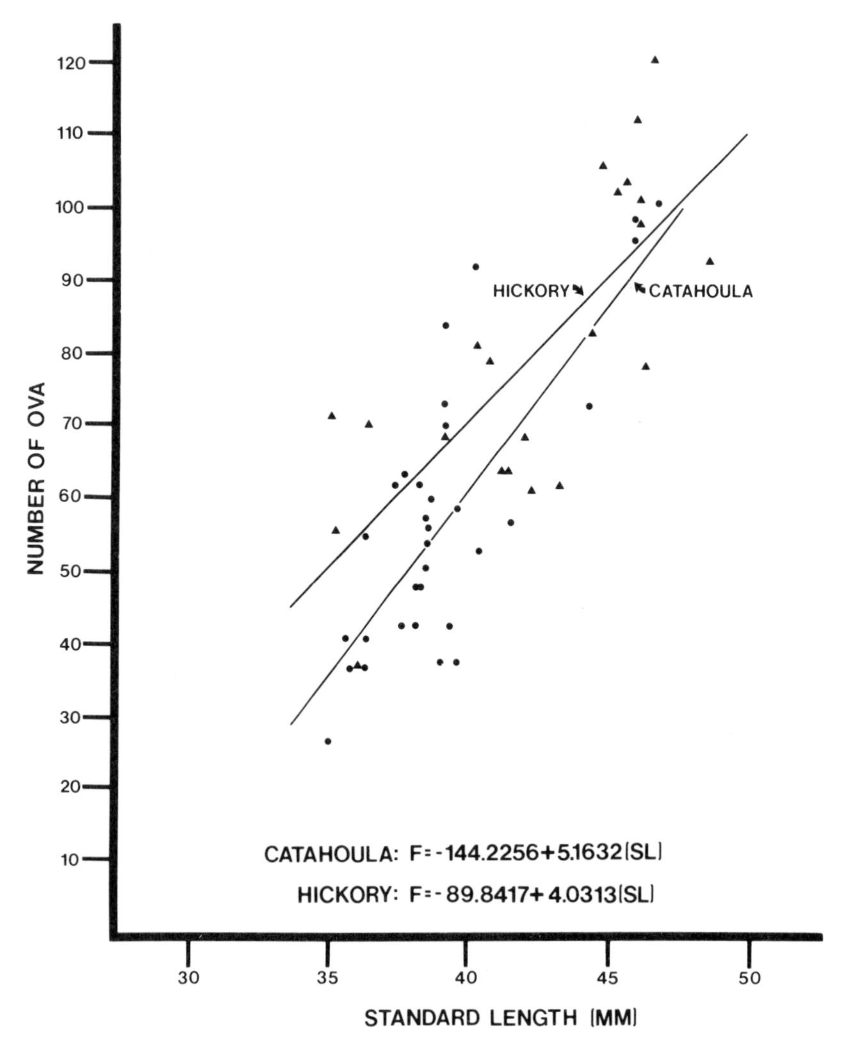

Fig. 2. Relationship between number of mature ova and SL for *Ammocrypta beani* in Catahoula Creek (1971, dots) and Hickory Creek (1976, triangles) Hancock County, Mississippi.

ber and body size were adequately defined by either model. Logarithmic transformation of both variables stabilizes the variance along the regression line (Bagenal 1967) and is important for use in prediction, therefore, we provide these equations herein. For Catahoula Creek (1971) the relationship is ($r^2 = 0.747$, $p<0.01$): $\log_{10}F = -3.6342 + 3.3770(\log_{10}SL)$. The relationship for Hickory Creek is ($r^2 = 0.739$, $p<0.01$): $\log_{10}F = 1.6287 + 2.1669(\log_{10}SL)$.

Size of mature ova

There was no significant correlation between mean mature ovum diameter and SL for Catahoula Creek ($r = -0.125$, $n = 12$, $p>0.05$) or Hickory Creek ($r = 0.193$, $n = 22$, $p>0.05$). Individual mature ova in females from Catahoula Creek ranged from 0.70–1.05 mm diameter, averaging 0.90 mm. Mature ova in females from Hickory Creek averaged 0.93 mm and ranged from 0.60–1.15 mm diameter. Student's t-test (Sokal & Rohlf 1969) showed that there was no significant difference in mean mature ovum diameter for the two localities ($t = 0.5245$, $df = 32$, $p>0.05$). The combined mean mature ovum diameter was 0.92 mm. Williams (1975) reported that mature ova in *A. beani* were about 1 mm in diameter.

Table 2. Results of analyses of covariance comparing ovum number-body size regressions for *Ammocrypta beani* from Catahoula Creek (1971) and Hickory Creek (1976), Hancock County, Mississippi.

Source of variance	df	Mean square
Within	50	173.57
Slopes	1	184.30
Pooled	51	173.79
Adjusted means	1	762.74
Total	52	185.11

$F_{slope} = 184.30/173.57 = 1.0618^{ns}$
$F_{elev} = 762.74/173.79 = 4.3889^{*}$

ns = non-significant, * = significance at 0.05 level.

Ovarian weight-somatic weight ratio

No significant correlation existed between ovarian weight-somatic weight ratio and SL for Catahoula Creek ($r = -0.199$, $n = 32$, $p>0.05$) or Hickory Creek ($r = -0.128$, $n = 22$, $p>0.05$). Females from Hickory Creek had an average ovarian weight-somatic weight ratio of 9.1% (range = 4.5–14.4%). Catahoula Creek females had an average ratio of 8.5% (range = 4.4–14.0%). Student's t-test showed no significant difference in the mean ovarian weight-somatic weight ratio for the two localities ($t = 0.9356$, $df = 52$, $p>0.05$). The combined mean ovarian weight-somatic weight ratio was 8.8%.

Size at sexual maturity

Collections of *A. beani* taken from Catahoula Creek during the middle of the 1971 reproductive season (May, June, July) were analyzed to determine size at sexual maturity. Most males (64%, n = 11) and females (92%, n = 25) 32 mm in length were mature. All specimens 34 mm in length (26♀♀, 15♂♂) were mature. The largest immature male was 33.2 mm SL; the largest immature female was 32.8 mm SL.

Age and growth

Collections of *A. beani* are typically small (Williams 1975, and personal communication). By comparison, our collections are consistently large. Nevertheless we did not think our collections were adequate for length frequency analysis based on all months of the year. Only collecting periods for which large numbers of specimens were available have been analyzed (Fig. 3).

Young of the year (26–32 mm SL) were present in early August. Adults ≥34 mm SL formed a bimodal group. By early November, these two

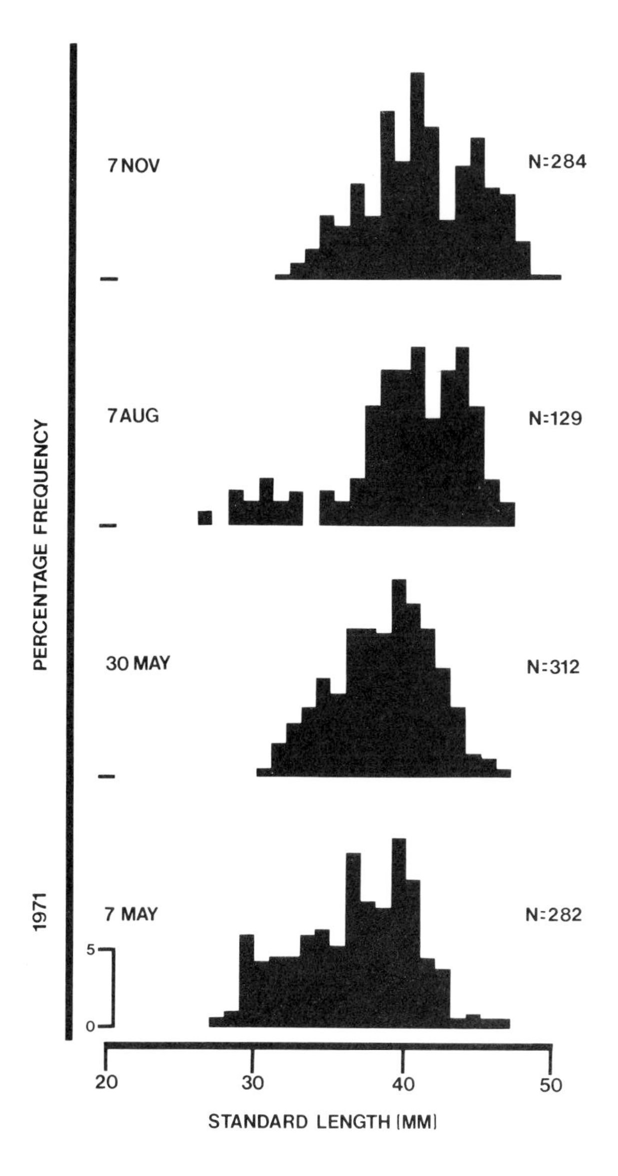

Fig. 3. Length-frequency histograms for collections of *Ammocrypta beani* from Catahoula Creek, Hancock County, Mississippi, in 1971.

groups were indistinguishable. Each of the collections from May consisted of a uni-modal group of individuals, presumably composed of two successive year-classes which survived the winter months. Therefore, it appears that the maximum lifespan is about 2 years. Based on these data and data on size at sexual maturity, some individuals may begin reproducing within their first season of growth; however, most *A. beani* apparently mature at or about one year during their second season of growth.

Discussion

Few detailed life history studies have been published on the darters of the southern United States, particularly those inhabiting states along the Gulf Coastal Plain. There is also a paucity of detailed information on members of the genus *Ammocrypta.* This makes it difficult to generalize about the life history of *Ammocrypta beani.*

Ammocrypta beani in southeastern Mississippi has a reproductive season that lasts about six months. This is considerably longer than that previously reported for *A. beani* (Williams 1975). Thus the reproductive seasons of the closely related *A. bifascia* and southern populations of other species in the genus *Ammocrypta* may also be longer than reported by Williams (1975). The reproductive season of *A. beani* is also considerably longer than that reported for most other species of darters in the southern United States (e.g. *Etheostoma boschungi,* Boschung personal communication, *Etheostoma coosae,* O'Neil 1982, *Etheostoma perlongum,* Lindquist et al. 1981, *Etheostoma proeliare,* Burr & Page 1978, *Etheostoma radiosum,* Scalet 1973, *Percina macrolepida,* Stevenson 1971, *Percina nigrofasciata,* Mathur 1973). The exceptions to this generalization are species which inhabit thermally stable or stabilized spring-fed streams: *Etheostoma lepidum* and *Etheostoma spectabile* (Hubbs & Strawn 1957, Hubbs et al. 1968) and *Etheostoma fonticola* (Schenck & Whiteside 1977). In these cases, thermal stability was related to prolonged reproductive seasons.

Although the long reproductive season of *A. beani* is probably facilitated by the long period of relatively high temperatures (Fig. 1) and long daylengths (U.S. Naval Observatory, Washington, D.C.) in the region, it may be adaptive to a variable environment (Starrett 1951, Tanyolac 1973, Wallace 1973, Heins & Bresnick 1975, Heins et al. 1980, Heins 1981). Heins & Clemmer (1976) discussed this life history trait for the cyprinid *Notropis longirostris* from the Jourdan River drainage. By distributing spawning over a long period of time, the chance of losing a large portion of the annual recruitment may be reduced.

Balon et al. (1977) listed *Ammocrypta* spp. as belonging to the psammophilous reproductive guild of non-guarding fishes. They consider this guild to be relatively r-selected in comparison to the entire array of guilds delineated by Balon (1975). Our data support this alignment in that *A. beani* is a relatively small, short-lived species with a prolonged reproductive season; and maturity is early, occurring at about one year.

Acknowledgements

Gary Miller loaned us the specimens of *Ammocrypta beani* in the Mississippi State University Collection of Fishes. The Department of Biology and the Dean of Millsaps College, provided financial support. Marjorie Canada, Millsaps College, and Terri Bensel and Marcia Sytsma, Tulane University, typed the manuscripts.

References cited

Bagenal, T.B. 1967. A short review of fish fecundity. pp. 89–111. *In*: S.D. Gerking (ed.) The Biological Basis of Freshwater Fish Production, Blackwell Scientific Publications, Oxford.

Bailey, R.M., H.E. Winn & C.L. Smith. 1954. Fishes of the Escambia River, Alabama and Florida, with ecological and taxonomic notes. Proc. Acad. Nat. Sci. Phila. 106: 109–164.

Bailey, R.M. & W.A. Gosline. 1955. Variation and systematic significance of vertebral counts in American fishes of the family Percidae. Misc. Publ. Mus. Zool. Univ. Mich. 93: 1–44.

Balon, E.K. 1975. Reproductive guilds of fishes: a proposal and definition. J. Fish. Res. Board Can. 32: 821–864.

Balon, E.K., W.T. Momot & H.A. Regier. 1977. Reproductive

guilds of percids: results of the paleogeographical history and ecological succession. J. Fish. Res. Board Can. 34: 1910–1921.

Burr, B.M. & L.M. Page. 1978. The life history of the cypress darter, *Etheostoma proeliare,* in Max Creek, Illinois. Ill. Nat. Hist. Surv. Biol. Notes 106: 1–15.

Heins, D.C. 1979. A comparative life history of a closely related group of minnows (*Notropis,* Cyprinidae) inhabiting streams of the Gulf Coastal Plain. Ph.D. Thesis, Tulane University, New Orleans. 161 pp.

Heins, D.C. 1981. Life history pattern of *Notropis sabinae* (Pisces: Cyprinidae) in the lower Sabine River drainage of Lousiana and Texas. Tulane Stud. Zool. and Bot. 22: 67–84.

Heins, D.C. & G.I. Bresnick. 1975. The ecological life history of the cherryfin shiner, *Notropis rosiepinnis.* Trans. Amer. Fish. Soc. 104: 516–523.

Heins, D.C. & G.H. Clemmer. 1976. Ecology, foods and feeding of the longnose shiner, *Notropis longirostris* (Hay), in Mississippi. Amer. Midl. Nat. 94: 284–295.

Heins, D.C. & G.H. Clemmer. 1976. The reproductive biology, age and growth of the North American cyprinid, *Notropis longirostris* (Hay). J. Fish Biol. 8: 365–379.

Heins, D.C., G.E. Gunning & J.D. Williams. 1980. Reproduction and population structure of an undescribed species of *Notropis* (Pisces: Cyprinidae) from the Mobile Bay drainage, Alabama. Copeia 1980: 822–830.

Hickling, C.F. & E. Rutenberg. 1936. The ovary as an indicator of the spawning period in fishes. J. Mar. Biol. Assn. U.K. 21: 311–318.

Hubbs, C. & K. Strawn. 1957. The effects of light and temperature on the fecundity of the greenthroat darter, *Etheostoma lepidum.* Ecology 38: 596–602.

Hubbs, C., M.M. Stevenson & A.E. Peden. 1968. Fecundity and egg size in two central Texas darter populations. Southwest. Nat. 13: 301–324.

Lindquist, D.G., J.R. Shute & P.W. Shute. 1981. Spawning and nesting behavior of the waccamaw darter, *Etheostoma perlongum.* Env. Biol. Fish. 6: 177–191.

MacGregor, J.S. 1970. Fecundity, multiple spawning, and description of the gonads in *Sebastodes.* U.S. Fish. Wildl. Serv. Spec. Sci. Rept. Fish. No. 596.

Mathur, D. 1973. Some aspects of life history of the blackbanded darter, *Percina nigrofasciata* (Agassiz), in Halawakee Creek, Alabama. Amer. Midl. Nat. 89: 381–393.

O'Neil, P.E. 1982. Life history of *Etheostoma coosae* (Pisces: Percidae) in Barbaree Creek, Alabama. Tulane Stud. Zool. and Bot. 23: 75–83.

Page, L.M. 1981. The genera and subgenera of darters (Percidae, Etheostomatini). Occ. Pap. Mus. Nat. Hist. Univ. Kansas 90: 1–69.

Prabhu, M.S. 1956. Maturation of intro-ovarian eggs and spawning periodicities in some fishes. Indian J. Fish. 3: 59–90.

Qasim, S.Z. & A. Qayyum. 1961. Spawning frequencies and breeding seasons of some freshwater fishes, with special reference to those occurring in the plains of northern India. Indian J. Fish. 8: 24–43.

Scalet, C.G. 1973. Reproduction of the orangebelly darter, *Etheostoma radiosum cyanorum* (Osteichthyes: Percidae). Amer. Midl. Nat. 89: 156–165.

Schenck, J.R. & B.G. Whiteside. 1977. Reproduction, fecundity, sexual dimorphism and sex ratio of *Etheostoma fonticola* (Osteichthyes: Percidae). Amer. Midl. Nat. 98: 365–375.

Snedecor, G.W. & W.G. Cochran. 1967. Statistical methods. Iowa State Univ. Press, Ames. 593 pp.

Sokal, R.R. & F.J. Rohlf. 1969. Biometry. Freeman, San Francisco. 776 pp.

Starrett, W.C. 1951. Some factors affecting the abundance of minnows in the Des Moines River, Iowa. Ecology 32: 13–27.

Stevenson, M.M. 1971. *Percina macrolepida* (Pisces, Percidae, Etheostomatini), a new percid fish of the subgenus *Percina* from Texas. Southwest. Nat. 16: 65–83.

Tanyolac, J. 1973. Morphometric variation and life history of the cyprinid fish *Notropis stramineus* (Cope). Occ. Pap. Mus. Nat. Hist. Univ. Kansas 12: 1–28.

Wallace, D.C. 1973. Reproduction of the silverjaw minnow *Ericymba buccata* Cope. Trans. Amer. Fish. Soc. 102: 786–793.

Williams, J.D. 1975. Systematics of the percid fishes of the subgenus *Ammocrypta,* genus *Ammocrypta,* with descriptions of two new species. Bull. Alabama Mus. Nat. Hist. 1: 1–56.

Received 14.6.1982 *Accepted 21.7.1983*

The dusky darter *Percina sciera* (top specimen, by permission of University of Illinois Press, from P.W. Smith 'The Fishes of Illinois') and the slenderhead darter *Percina phoxocephala*.

Life history of *Etheostoma caeruleum* (Pisces: Percidae) in Bayou Sara, Louisiana and Mississippi

James M. Grady[1] & Henry L. Bart, Jr.[2]
Department of Biological Sciences, University of New Orleans New Orleans, LA 70148, U.S.A.

Keywords: Fecundity, Reproduction, Food habits, Spawning, Parasites, Diet, Sex ratio, Rainbow darter, Habitat

Synopsis

The life history of *Etheostoma caeruleum* was studied from collections made between September 1976 and July 1981 in Bayou Sara, a lower Mississippi River tributary. The principal habitat of rainbow darters was shallow gravel riffles or runs over firm substrates. In Bayou Sara, the reproductive season extended from late March through April and possibly into June. Males and females were sexually mature in their first year. Counts of unovulated mature ova ranged from 17–125 ($\bar{X} = 66$). Rainbow darters reached a maximum age of 37 months and a maximum size of 50 mm. Individuals reached an average of 34 mm in the first year and 41 mm after two years. A significantly higher number of males than females was present in the 1+ and 2+ year classes. Chironomids, siphlonurids, and simuliids comprised the bulk of the diet. Simuliid and hydropsychid consumption increased with increased size.

Introduction

Etheostoma caeruleum is one of several widely distributed members of the subgenus *Oligocephalus* (Page 1981a). Excluding *E. caeruleum,* ecological or life history information has been presented for only four of the fifteen members of *Oligocephalus*: *E. asprigene* (Grady & Cummings 1982), *E. lepidum* (Strawn 1955, 1956, Hubbs & Strawn 1957, Hubbs & Delco 1960, Hubbs 1961, Hubbs & Martin 1965, Hubbs et al. 1968), *E. radiosum* (Scalet 1971, 1972, 1973a, 1973b, 1974) and *E. spectabile* (Winn 1958a, 1958b, Hubbs & Armstrong 1962, Hubbs et al. 1968, Marsh 1980). Published information on *E. caeruleum* includes studies on reproductive biology by Reeves (1907 – actually a combined study of *E. caeruleum* and *E. spectabile*), Winn (1957, 1958a, 1958b) and Lutterbie (1979), egg and larval descriptions by Fish (1932) and Cooper (1979), age and growth by Lutterbie (1979), and feeding habits by Turner (1921) and Adamson & Wissing (1977). This study was conducted to summarize the ecology and life history of *E. caeruleum* in Bayou Sara, a tributary to the lower Mississippi River.

[1] Present address: Department of Zoology, Southern Illinois University, Carbondale, IL 62901, U.S.A.
[2] Present address: Department of Zoology, University of Oklahoma, Norman, OK 73019, U.S.A.

Study area

Etheostoma caeruleum was studied from collections made in upstream Bayou Sara, West Feliciana Parish, Louisiana and Wilkinson County, Mississippi. In its headwaters, Bayou Sara is comprised of a series of shallow (7–27 cm) and narrow

David G. Lindquist & Lawrence M. Page (ed.), Environmental biology of darters. ISBN 90-6193-506-7
 Developments in EBF 4.

(0.75–2.5 m) spring fed tributaries. Extensive shallow runs over white sand intermixed with small gravel are typical in these tributaries. Interspersed between runs are clay bottomed pools and occasional sheet riffles over patches cf mixed gravel.

Measured physicochemical characteristics of the headwaters of Bayou Sara include: total alkalinity, 35–37 mg l^{-1}; free CO_2, 1.4–9.0 mg l^{-1}; dissolved oxygen, 9.7–13.2 mg l^{-1}; temperature, 4.3–28.0° C; current, 0.21–0.38 m sec^{-1}; conductivity, 53–75 mhos cm^{-2}; total solids, 48.7–91.0 mg l^{-1}; suspended solids, 14.4–24.5 mg l^{-1}; turbidity, 0.3–2.6 mg l^{-1}.

Methods

E. caeruleum was collected at about 2-month intervals from various sites in upstream Bayou Sara from September 1976 through April 1979. Supplemental collections to obtain additional reproductive data were made in the spring of 1978 and 1979 and summer 1981. Collections were made with 4.8 mm or 3.2 mm mesh seines or dip nets. All available habitat types were sampled each month to determine habitat preference of juveniles and adults. A total of 600 specimens taken in 33 collections were fixed in 10% formalin and stored in 40% isopropyl alcohol. Specimens are deposited at the University of New Orleans. Potential predators taken with *E. caeruleum* were preserved for stomach content analysis.

Preserved specimens were measured to the nearest 0.1 mm standard length, sexed, aged, and examined for incidence of ectoparasites. Aging to year class was accomplished through analysis of scales removed from the dorsum just below the insertions of the dorsal fins and from the caudal peduncle. Length frequency histograms also were used in aging specimens. Aging to month was accomplished by using April, the month of greatest breeding activity, as month zero (Page 1974).

A minimum of ten individuals were dissected each month except May for stomach content analysis and endoparasite identification. Consistently high water levels prevented adequate sampling in May. Individual prey items were identified and counted to obtain the percentage of each taxon in the diet and the percentage of stomachs in which each taxon occurred. A total of 157 stomachs were examined.

Reproductive condition was assessed by overall body coloration, development of the genital papilla, gross examination of the gonads, and determination of gonad weights. Gonads of a number of males and females from each collection were removed, air dried, and weighed to the nearest 0.001 g; a gonadosomatic index (GSI) was recorded as the ratio of gonad weight (× 10,000 for ovaries, × 1,000 for testes) to adjusted body weight (specimen minus the liver, intestines, and stomach). Mature ova in 19 females were counted and measured; mean diameters were estimated by averaging the largest and smallest diameters.

Results and discussion

Habitat

Etheostoma caeruleum juveniles and adults were consistently taken over firm gravel riffles and runs. Juveniles were most frequently encountered in shallow riffles and runs or along the banks in deeper riffles. Rainbow darters were rarely found over the numerous gravel patches overlying soft sand. A slight seasonal shift in habitat use was observed. During winter months when water temperatures fell below 10° C, smaller individuals congregated in pools at the base of riffles or in the deeper water between sand-gravel runs. Only the largest adults remained in riffles.

Etheostoma caeruleum was the most abundant percid in Bayou Sara. *Etheostoma whipplei* and *Percina ouachitae,* the only other percids taken with *E. caeruleum,* were much less abundant and also exhibited distinct habitat preferences. *Percina ouachitae* was most frequently taken in deep runs over hard packed clay substrates. *Etheostoma whipplei* was commonly associated with aquatic vegetation in small, spring fed tributaries in the extreme headwaters of Bayou Sara. Other fishes taken with *E. caeruleum* in Bayou Sara in order of abundance were *Notropsis venustus, N. chryso-*

cephalus, Nocomis leptocephalus, Notropis longirostris, Pimephales notatus, Fundulus olivaceus, Lepomis megalotis, Lepomis macrochirus, Notropsis volucellus, Semotilus atromaculatus, Erimyzon oblongus, Gambusia affinis, Ictalurus natalis, Pimephales vigilax, and *Carpiodes carpio.*

Demography

Seasonal density estimates were taken at one station in Bayou Sara by repeatedly seining an area until no more individuals were taken. The density of *E. caeruleum* in firm gravel sheet riffles, the preferred habitat, was 4.3 individuals m^{-2} on 29 December 1978, 8.9 individuals m^{-2} on 12 March 1978, 0.9 individuals m^{-2} on 24 June 1978, and 7.7 individuals m^{-2} on 12 October 1978. The relatively low density observed in June 1978 was apparently influenced by extremely low water levels which may have forced individuals into pools or deeper riffles in other areas. The December estimate is consistent with the movement of juveniles into pools. The mean density of *E. caeruleum* in sheet riffles, 5.54 m^{-2}, is higher than the 2.66 m^{-2} reported for *E. radiosum* (Scalet 1973a) and other riffle inhabiting species. Density estimates for *E. caeruleum* more closely approximate those of *E. smithi* (Page & Burr 1976), a pool inhabiting species, and *E. microperca* (Burr & Page 1979), an inhabitant of quiet waters with abundant submerged vegetation.

Reproduction

The spectacular breeding colors of male *E. caeruleum* were described intially by Storer (1845) and in detail by Jordan & Evermann (1896). Prior to the development of spawning colors, juvenile males possess the relatively drab coloration typical of females. In Bayou Sara, males, especially those greater than 40 mm SL, developed breeding coloration as early as late October following the first dramatic temperature drop. Coloration faded considerably during the winter but did not disappear completely. Breeding colors intensified again in late January and males remained brightly colored through April.

In addition to the bright color changes, male *E. caeruleum* undergo an overall increase in melanophore concentration in the pectoral fins, the breast and abdominal regions, and on the genital papilla prior to the breeding season. The female shows a slight increase in pigmentation in all these areas with the exception of the genital papilla which remains without pigment throughout the year.

The genital papilla of nonbreeding male *E. caeruleum* is a small triloboded, flap-like structure covering the genital pore. In late December, it begins to enlarge without noticeable conformational changes and reaches its maximum development in late February. The genital papillae of nonbreeding females are essentially identical to those of nonbreeding males. Enlargement and conformational changes in the female genital papilla commence in late November and culminate in mid March in an elongate tubular structure with a series of finger-like projections at the tip (Fig. 1). In spite of differences in reproductive habits among species of *Oligocephalus,* the shape and configuration of the genital papilla of *E. caeruleum* does not differ significantly from that of *E. asprigene* (personal observation), *E. ditrema* (Ramsey & Suttkus 1965), *E. radiosum* (Scalet 1973b), or *E. swaini* (Collette & Yerger 1962).

All spring-collected males (N = 89) were brightly colored and had enlarged genital papillae and enlarged, white, granular testes. These specimens were 1 to 2 years old and ranged from 29–50 mm SL ($\bar{x}$ = 40 mm). Sexual maturity of males in the first year following hatching is common in small, short-lived percids; e.g., *E. proeliare* (Burr & Page 1978), *E. microperca* (Burr & Page 1979), *E. striatulum* (Page 1980), *E. simoterum* (Page & Mayden 1981), and has been reported in several larger, long-lived species including *E. flabellare* (Lake 1936), *E. blennioides* (Fahy 1954), and *E. radiosum* (Scalet 1973b). A 29 mm male with enlarged testes and free flowing milt is smaller than that found in several other darters, e.g., *E. olmstedi* and *E. longimanum* (Raney & Lachner 1943), *E. variatum* and *E. zonale* (Lachner et al. 1950), *E. squamiceps* (Page 1974), and *E. olivaceum* (Page 1980).

June-collected males demonstrated various

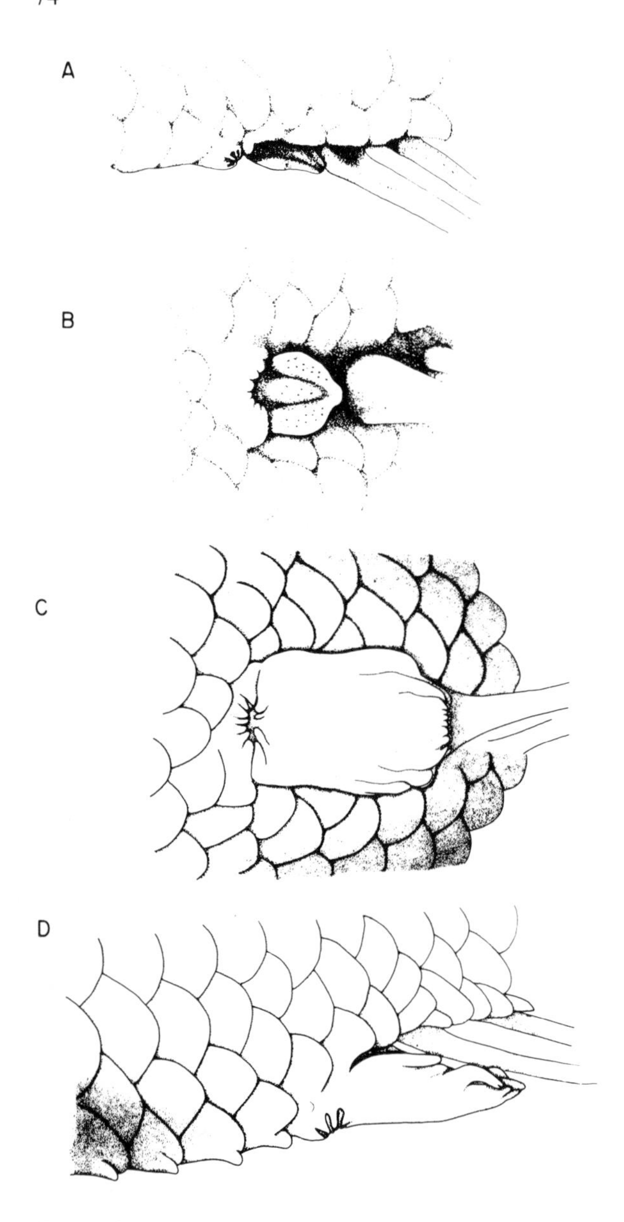

Fig. 1. Genital papillae of breeding *Etheostoma caeruleum*. A and B are lateral and ventral views, respectively, of a male; C and D are ventral and lateral views, respectively, of a female.

stages of testicular regression. Testes of all July-collected males were reduced to transparent filamentous structures and remained in this state until late October. At that time testes of larger males began to increase and became opaque. Enlargement and color changes of the testes were apparent in all November-collected specimens. The relationship between the ratio of testes-to-adjusted body weight (X) and month (Y) with November = 1 and April = 6 was $Y = 0.2663 - 0.0899X + 0.0097X^2 - 0.0003X^3$, with $r = 0.80$. Testes of all males collected between May and October were too small to allow accurate weight determination (Fig. 2).

Gross examination of ovaries and GSI values indicated that spawning of *E. caeruleum* in Bayou Sara occurs as early as late March and continues through May. Although considerable variation exists within and between populations, a late spring to early summer spawning season has been reported throughout the range of *E. caeruleum* (Reeves 1907, Winn 1958b, Lutterbie 1979). Published information on the reproductive season of other species of *Oligocephalus* indicates a spring-summer spawning for *E. australe* (Meek 1904, Page 1981b), *E. grahami* (Harrel 1980), *E. spectabile* (Winn 1958b), *E. ditrema* (Ramsey & Suttkus 1965), *E. radiosum* (Scalet 1973b), *E. swaini* (Burr & Mayden 1979), and *E. asprigene* (Grady & Cummings 1982). Notable exceptions include *E. lepidum* and central Texas populations of *E. spectabile* which spawn from October to May and November to April, respectively (Hubbs et al. 1968).

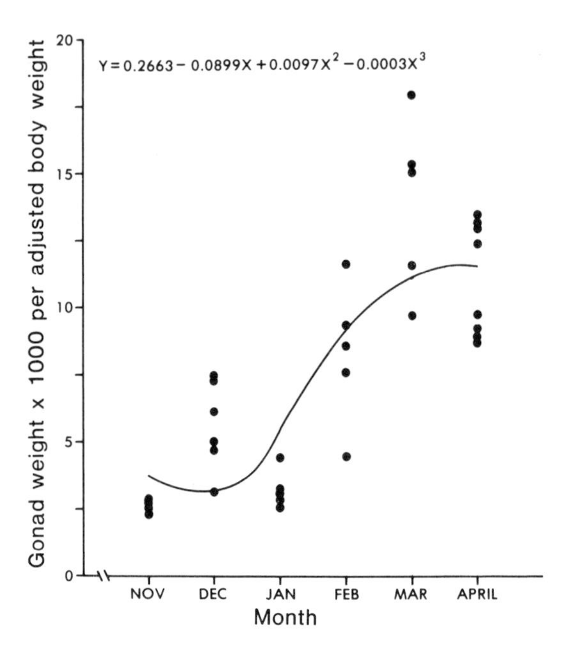

Fig. 2. Monthly variations in testicular weight relative to adjusted body weight of 33 *Etheostoma caeruleum*. Testes of males collected between July and November were too small to allow accurate weight determination.

The presence of 45 mature ova in a 35 mm female collected on 24 June 1978 suggests an extended spawning season. Winn (1958b) attributed yearly variation in spawning season in Michigan populations to prolonged cold temperatures and/or increased turbidity due to heavy rainfall. Neither condition prevailed in Bayou Sara during the spring of 1978 suggesting that the breeding season may be affected by other environmental factors.

All late March to April-collected females possessed mature ova. Individuals were 1 to 2 years old and ranged from 28–46 mm ($\bar{x} = 37$ mm). Mature ova were spherical to subspherical, dark yellow to orange or translucent with one or two large oil droplets and several smaller ones, and were deeply indented on one side. This indentation was present only in fully mature ova. Indented eggs have been described from several species in the subgenera *Boleichthyes* and *Catonotus* (Burr & Page 1978, 1979, Burr & Ellinger 1980, Page 1980, Page & Mayden 1981) but have not been reported previously in *Oligocephalus*. Fully mature ova were 1.2–1.8 mm in diameter. Reeves (1907) reported 1.5 mm ova in an Ohio population of *E. caeruleum* while Winn (1958b) found slightly larger eggs (1.6–1.8 mm) in Michigan populations. Cooper (1979) observed 1.7–1.9 mm ova in Elk Creek, Pennsylvania. Overall, mature ova of *E. caeruleum* and other egg burying (Page 1983) or brood hiding lithophil (Balon 1975) species of *Oligocephalus* (e.g., *E. radiosum,* 1.2–1.5 mm – Scalet 1973b; *E. spectabile,* 1.2–1.5 mm – Hubbs et al. 1968) are larger than those of egg attaching members of *Oligocephalus* and other subgenera (e.g. *E. ditrema,* 1.1–1.2 mm – Ramsey & Suttkus 1965; *E. lepidum,* 1.3 mm – Hubbs et al. 1968; *E. asprigene,* 1.0–1.4 mm – Grady & Cummings 1982).

The number of mature ova in 19 females collected in March, April, and June ranged from 17–125 ($\bar{x} = 66$) (Table 1). For March-collected females ($N = 11$), there was a positive correlation between the number of mature ova produced by a female and standard length ($r = 0.59$, $p<0.05$) but no significant correlation between the number of mature ova and adjusted body weight ($r = 0.56$, $p>0.05$). A significant negative correlation existed between the number of mature ova and standard length ($r = -0.65$, $p<0.05$) and the number of mature ova and adjusted body weight ($r = 0.77$, $p<0.05$) of April-collected females ($N = 7$). The

Table 1. Relationship between size of female, ovary weight, and number of mature ova of *Etheostoma caeruleum.*

Month of collection	SL, mm	Adjusted body weight, g	Ovary weight, g	Number of mature (orange or translucent) ova
March	28	0.27	0.02	17
March	30	0.34	0.04	33
March	36	0.55	0.11	60
March	36	0.68	0.10	72
March	36	0.64	0.11	125
March	37	0.68	0.10	60
March	37	0.68	0.18	79
March	38	0.80	0.19	59
March	41	0.86	0.21	49
March	42	1.01	0.12	75
March	42	0.92	0.26	116
April	34	0.47	0.12	93
April	36	0.64	0.10	43
April	37	0.59	0.16	85
April	37	0.60	0.11	64
April	38	0.66	0.14	87
April	40	0.81	0.07	68
April	43	1.01	0.12	22
June	37	0.66	0.05	45

change of slope in these relationships between months indicated that ova development and spawning occurs first in the larger females, and that small individuals mature and spawn later in the season. Fahy (1954) described a similar situation for *E. blennioides* in New York.

The appearance of differentiated ova and subsequent ovarian enlargement began in larger females (35 mm SL or greater) in late October. Between July and October ovaries of female *E. caeruleum* remained small and translucent with only undifferentiated ova (clear) present. Enlarged ovaries and differentiated ova were present in all November-collected individuals. By late January yellow ova were present. Ova size and coloration increased with fully mature, dark yellow to orange translucent ova first appearing in March.

The relationship between GSI (Y) and month (X) with May = 1 and April = 12 was log Y = $3.9964 - 1.1810X + 0.1681X^2 - 0.0062X^3$, with r = 0.92. The proportionally largest ovaries (GSI = 2826) were found in a 42.1 mm female collected on 31 March 1979 (Fig. 3).

Field observations to study the spawning behavior of *E. caeruleum* were made on several occasions in March 1978. Aggressive territorial and display behaviors were observed on these occasions, but spawning was not. Males established small roving territories in sheet riffles and gravel bottomed runs. Prior to the spawning season no territorial behavior was observed in either sex. Females remained at the periphery of the riffles and runs and were most abundant in shallow pools at the head and base of these riffles. Females entered the riffles periodically and were immediately pursued and courted by one or more males. Spawning did not occur and females returned to peripheral areas. Our observations agree closely with descriptions of prespawning behavior in *E. caeruleum* by Reeves (1907) and Winn (1957, 1958a). *Etheostoma radiosum* and *E. spectabile,* egg burying species of *Oligocephalus,* exhibit similar prespawning behavior (Winn 1957, 1958b, Scalet 1937b).

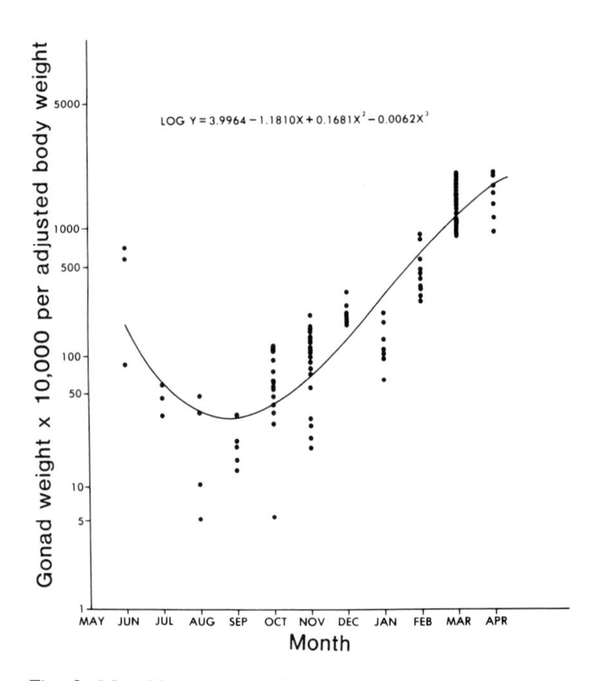

Fig. 3. Monthly variations in ovarian weight relative to adjusted body weight of 124 *Etheostoma caeruleum.*

Age and growth

Since individuals less than 20 mm were not taken during this study there is no information on the early development of *E. caeruleum* in Bayou Sara. In a description of embryos and larvae of rainbow darters from Elk Creek, Pennsylvania, Cooper (1979) reported an average length of 6.0 mm for fish at hatching. The smallest individual from Bayou Sara was a 23 mm female taken in May 1978. At 23 mm, squamation, pigmentation and head canal and lateral line formation were essentially complete. Eleven individuals collected in June and July and ranging from 23 to 27 mm exhibited a similar state of development.

In Bayou Sara, *E. caeruleum* reached an average of 34 mm (range = 31–36 mm) in the first year of growth (Fig. 4). At 12 months males averaged 33.9 mm while females averaged 34.4 mm (Fig. 5). At two years (24 months) *E. caeruleum* averaged 40.5 mm (range = 38–45 mm) (Fig. 4); females were slightly larger than males, 41.0 and 40.1 mm, respectively (Fig. 5). The oldest individuals taken in Bayou Sara were 37 month-old males, 46.7 and 49.8 mm SL, collected in April 1978 (Fig. 4). The oldest females, 36 months old and 45.1 and 45.5 mm, were taken in March 1978. The largest male, 49.8 mm, was collected in April 1978; the largest

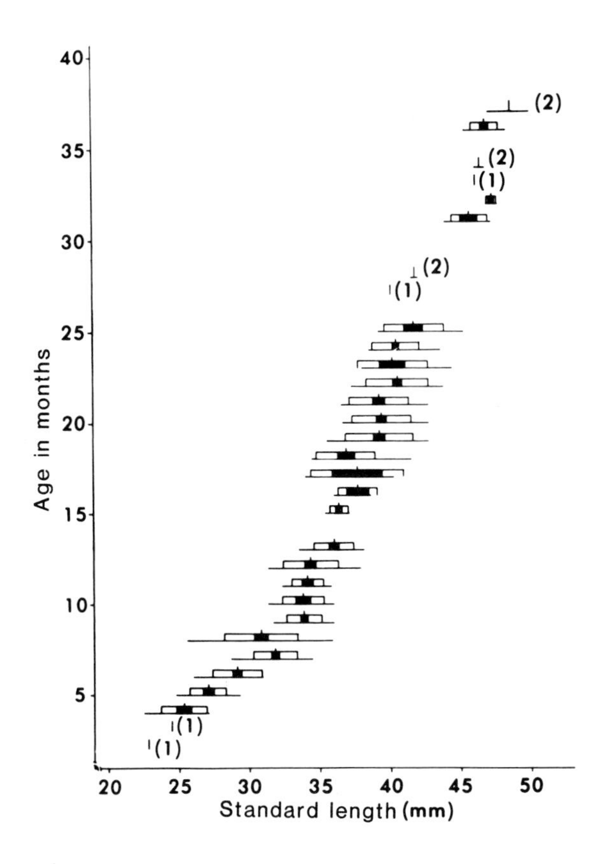

Fig. 4. Length age relationship of *Etheostoma caeruleum* in Bayou Sara. A total of 600 specimens is represented. Solid line equals range; vertical bar equals mean; solid rectangle equals 95% confidence interval; open rectangle equals one standard deviation on either side of the mean.

female was a 46.8 mm specimen taken in October 1978. Lutterbie (1979) indicated that Wisconsin populations of *E. caeruleum* reach 43.3 mm total length in their first year and 50.7 mm after two years. Male and female *E. caeruleum* grow at a rapid rate in the first six months of life reaching 50% of the first year's average growth in two months (Fig. 5). Length-age relationships for males and females were not significantly different ($p > 0.05$). Overall, growth rates slowed dramatically in the first winter and continued at a slower rate thereafter (Fig. 5).

Among the 600 *E. caeruleum* collected in Bayou Sara between September 1976 and July 1981, 55% of the individuals were in the 0+ year class, 39% were in the 1+ age class, and 7% were older than 2 years (Table 2). The 0+ age class (N = 231) was composed of nearly equal numbers of males and females. A significantly higher number of males (N = 132) than females (N = 99) was present in the 1+ age class ($p < 0.05$). This difference is most pronounced in the 2+ age class with 32 males and 10 females ($p < 0.005$). Overall, data from Bayou Sara indicates a higher survival for males. Assuming that our collections adequately represent age class proportions in the population, 83% of the first year males and 59% of the first year females survived to a second year. Similarly, 20% of the first-year males and only 6% of the first-year females survived to a third year. Overall, 11% of the young of the year *E. caeruleum* in Bayou Sara survived beyond two years.

Diet

A total of 17 prey taxa were identified in stomachs of 157 *E. caeruleum* (Table 3). Midge larvae (Chironomidae), black fly larvae (Simuliidae), and mayfly naiads (Siphlonuridae) predominated in the diet of all size classes (Fig. 6). These three taxa comprised 93% of the diet (total number of prey items ingested) of individuals 23–35 mm (SL) and 36–45 mm and 89% of the diet of specimens 46–50 mm. The smallest specimens, 23–35 mm, consumed a total of 11 taxa. Rainbow darters 36–45 mm had the most diverse diet with 13 taxa identified; the largest individuals, 46–50 mm, consumed the fewest taxa, 7. Turner (1921) found a similar

Table 2. Age-class distribution and survival of *Etheostoma caeruleum* in Bayou Sara. S_1 and S_2 represent survival from the 0+ and 1+ age classes, respectively.

Sex	Age class	Number	S_1	S_2
	0+	159	1.000	–
Males	1+	132	0.830	1.000
	2+	32	0.201	0.242
	0+	168	1.000	–
Females	1+	99	0.589	1.000
	2+	10	0.059	0.101
	0+	327	1.000	–
Combined sexes	1+	231	0.706	1.000
	2+	42	0.113	0.182

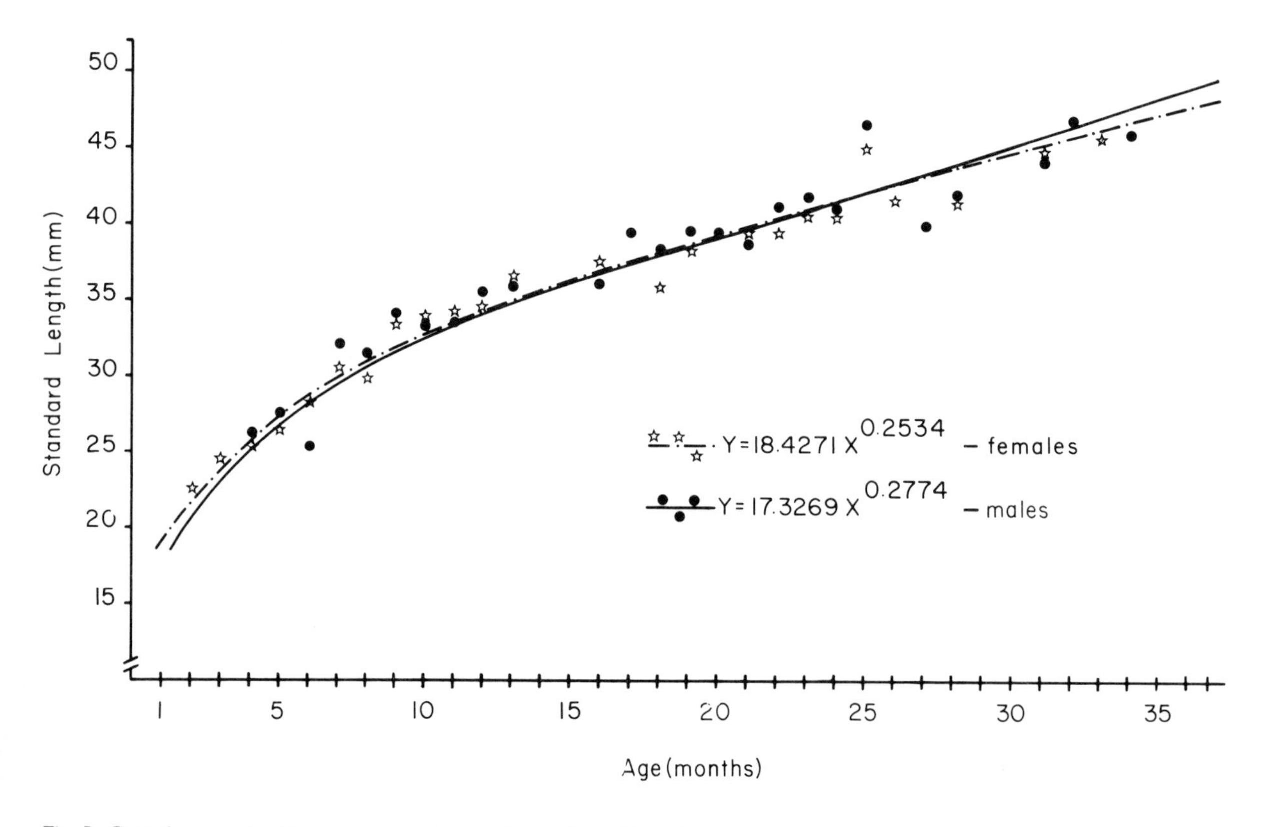

Fig. 5. Growth curves for *Esteostoma caeruleum* in Bayou Sara. Black dots represent sample means of males; stars represent sample means of females. Equations were not significantly different between sexes ($p>0.05$).

predominance of mayfly and midge larvae in the diets of Ohio populations of *E. caeruleum*. Adamson & Wissing (1977) reported a much higher incidence of caddisflies in one Ohio population.

Several size related trends in the diet of *E. caeruleum* were apparent (Fig. 6). There was a decreased utilization of midge larvae and water mites with increased size (SL). Conversely, caddisflies and black fly larvae consumption increased with increased size.

The predominance of midge and mayfly larvae in the diet of *E. caeruleum* was constant throughout the year (Table 3). Caddisflies and black flies represented a significant portion of the diet during several months, especially October and November. Diet diversity was highest in March (13 taxa), October, and November (8 taxa each). Food consumption was highest in June, July, and November and lowest in January and August.

Parasitism and predation

Fluke metacercariae were found on 4.6% of the specimens collected. Metacercariae were present on all size classes of darters. Infestation was heaviest in March, October, and December. Three of the 157 stomachs examined each contained one acanthocephalan, *Echinorhyncus* species, in February, March, and September.

Stomachs of forty-five specimens of five potential predator species taken in collections with *E. caeruleum* were examined. No evidence of predation on *E. caeruleum* was found in 8 *Semotilus atromaculatus* (113–157 mm), 11 *Lepomis macrochirus* (105–163 mm), 8 *Lepomis megalotis* (93–137 mm), 12 *Micropterus punctulatus* (128–223 mm), and 6 *Micropterus salmoides* (115–247 mm).

Table 3. Stomach contents of *Etheostoma caeruleum* from Bayou Sara by month of collection. Percent of stomach in which food organisms occurred is listed above and mean number of food organisms per stomach is listed in parentheses below. Sample sizes appear in parentheses below months.

Food organism	JAN (10)	FEB (10)	MAR (37)	APR (10)	MAY (1)	JUN (10)	JUL (10)	AUG (10)	SEP (10)	OCT (19)	NOV (20)	DEC (10)
Nematoda	....	....	....	....	....	....	10 (0.1)	....	....	....	....	....
Annelida												
Lumbricidae	....	....	6 (0.8)	....	....	....	....	....	....	....	....	....
Arachnida												
Acarina	....	60 (0.8)	6 (2.5)	....	....	....	10 (0.1)	....	20 (0.3)	5 (0.11)	25 (0.3)	20 (0.3)
Crustacea												
Cladocera	....	....	6 (0.2)	....	....	....	....	....	....	....	....	....
Ostracoda	....	....	3 (0.3)	....	....	....	....	....	....	....	....	....
Isopoda												
Asellidae	....	10 (0.1)	3 (0.6)	....	....	....	....	....	....	....	....	....
Amphipoda												
Gammaridae	....	....	3 (0.3)	....	....	....	....	....	....	....	....	....
Insecta												
Plecoptera	10 (0.1)	....	6 (0.2)	10 (0.4)	....	....	....	....	10 (0.1)	5 (0.1)	5 (0.1)	....
Ephemeroptera												
Siphlonuridae	50 (1.1)	....	46 (3.2)	40 (1.2)	100 (2)	70 (4.5)	90 (3.8)	20 (0.2)	20 (0.2)	47 (1.8)	40 (3.6)	....
Odonata												
Anisoptera	....	....	....	....	....	....	....	....	20 (0.2)	11 (0.1)	....	....
Trichoptera												
Hydropsychidae	10 (0.1)	20 (0.1)	8 (0.2)	....	....	....	20 (0.5)	....	....	21 (0.7)	55 (4.2)	....
Coleoptera	....	....	....	....	....	....	....	....	....	11 (0.1)	5 (0.1)	....
Diptera												
Ceratopogonidae	....	....	3 (0.1)	....	....	....	....	....	....	....	....	....
Chironomidae	70 (3.8)	100 (7.2)	68 (10.4)	40 (2.9)	100 (1)	80 (22.8)	100 (17.4)	....	50 (3.6)	53 (5.6)	95 (36.4)	36 (1.0)
Culicidae	....	....	....	....	....	....	....	....	....	....	10 (0.3)	....
Simuliidae	....	....	14 (0.2)	10 (0.1)	....	....	40 (1.1)	....	....	11 (0.8)	50 (19.6)	....
Teleost eggs	....	....	3 (.01)	....	....	....	....	....	....	....	....	....
Empty	10	0	51	0	0	20	0	90	50	32	10	30

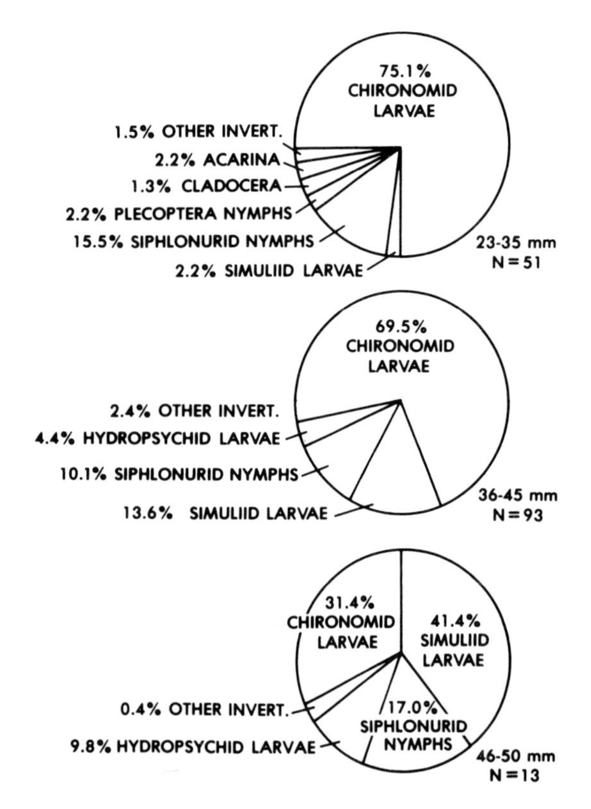

Fig. 6. Stomach contents (percent of total number of food items ingested) of *Etheostoma caeruleum* collected in Bayou Sara by size class of fish.

Acknowledgements

We are indebted to Millie Borel, Colleen Burrowes, Kimble D. Hobbs, R.W. Holzenthal, Lynne Junot, G.H. Laiche, C. Luquet, L. McCook, and F.L. Pezold for aid in collecting specimens and associated field work; to R.C. Cashner, Diana Stein, and M.M. Stevenson for reviewing the manuscript; to Willie Dillo and Susan Ray for assistance with the data analyses; and to B.M. Burr, K.S. Cummings, and T.E. Shepard for valuable discussions concerning darter life history strategies and numerous courtesies extended to J.M. Grady. This project was made possible in part through a grant from the Louisiana State University Foundation to J.M. Grady.

References cited

Adamson, S.W. & T.W. Wissing, 1977. Food habits and feeding periodicity of the rainbow, fantail, and banded darters in Four Mile Creek. Ohio J. Sci. 77: 164–169.

Balon, E.K. 1975. Reproductive guilds of fishes: a proposal and definition. J. Fish. Res. Board Can. 32: 821–864.

Burr, B.M & M.S. Ellinger. 1980. Distinctive egg morphology and its relationship to development in the percid fish *Etheostoma proeliare.* Copeia 1980: 556–559.

Burr, B.M. & R.L. Mayden. 1979. Records of fishes in Western Kentucky with additions to the known fauna. Trans. Ky. Acad. Sci. 40: 58–67.

Burr, B.M. & L.M. Page. 1978. The life history of the cypress darter, *Etheostoma proeliare,* in Max Creek, Illinois, Ill. Nat. Hist. Surv. Biol. Notes 106: 1–15.

Burr, B.M. & L.M. Page. 1979. The life history of the least darter, *Etheostoma microperca,* in the Iroquois River, Illinois. Ill. Nat. Hist. Surv. Biol. Notes 112: 1–15.

Collette, B.B. & R.W. Yerger. 1962. The American percid fishes of the subgenus *Villora.* Tulane Stud. Zool. 9: 213–230.

Cooper, J.E. 1979. Description of eggs and larvae of fantail (*Etheostoma flabellare*) and rainbow (*E. caeruleum*) darters from Lake Erie tributaries. Trans. Amer. Fish. Soc. 108: 46–56.

Fahy, W.E. 1954. The life history of the northern greenside darter, *Etheostoma blennioides blennioides* Rafinesque. J. Elisha Mitchell Sci. Soc. 70: 139–205.

Fish, M.P. 1932. Contributions to the early life history of sixty-two species of fishes from Lake Erie and its tributary waters. Bull. U.S. Bur. Fish. 49: 293–398.

Grady, J.M. & K.S. Cummings. 1982. Life history aspects of the mud darter, *Etheostoma asprigene,* in Lake Creek, Alexander Co., Illinois. Abstract 62 Ann. Meeting Amer. Soc. Ichthyol. and Herp., DeKalb, Illinois.

Harrel, H.L. 1980. *Etheostoma grahami* (Girard), Rio Grande darter. p. 652. *In*: D.S. Lee et al. Atlas of North American Freshwater Fishes, N.C. State Mus. Nat. Hist., Raleigh.

Hubbs, C. 1961. Differences in the incubation period of two populations of *Etheostoma lepidum.* Copeia 1961: 198–200.

Hubbs, C. & N.E. Armstrong. 1962. Developmental temperatures of Texas and Arkansas-Missouri *Etheostoma spectabile* (Percidae: Osteichthyes). Ecology 43: 742–744.

Hubbs, C. & E.A. Delco. 1960. Geographic variations in egg complement of *Etheostoma lepidum.* Tex. J. Sci. 12: 3–7.

Hubbs, C. & P.S. Martin. 1965. Effects of darkness on egg deposition by *Etheostoma lepidum* females. Southwest. Nat. 10: 302–306.

Hubbs, C. & K. Strawn. 1957. The effects of light and temperature on the fecundity of the greenthroat darter, *Etheostoma lepidum.* Ecology 38: 596–602.

Jordan, D.S. & B.W. Evermann. 1896. Fishes of North and Middle America. U.S. Natl. Mus. Bull. 47: 1–1240.

Lachner, E.A., E.F. Westlake & P.S. Handwerk. 1950. Studies on the biology of some percid fish from western Pennsylvania. Amer. Midl. Nat. 43: 92–111.

Lake, C.T. 1936. The life history of the fantailed darter, *Catonotus flabellaris flabellaris* (Rafinesque). Amer. Midl. Nat. 17: 816–830.

Lutterbie, G.W. 1979. Reproduction and age and growth in Wisconsin darters (Osteichthyes: Percidae). Univ. Wisconsin Mus. Nat. Hist. Rep. Fauna Flora Wis. 15: 1–44.

Marsh, E. 1980. The effects of temperature and photoperiod on the termination of spawning in the orangethroat darter (*Etheostoma spectabile*) in central Texas. Tex. J. Sci. 32: 129–142.

Meek, S.E. 1904. The freshwater fishes of Mexico north of the Isthmus of Tehuanepec. Field Columbian Mus. Publ. 93, Zool. Series 5: 1–252.

Page, L.M. 1974. The life history of the spottail darter, *Etheostoma squamiceps,* in Big Creek, Illinois, and Ferguson Creek, Kentucky. Ill. Nat. Hist. Surv. Biol. Notes 93: 1–15.

Page, L.M. 1980. The life histories of *Etheostoma olivaceum* and *Etheostoma striatulum,* two species of darters in central Tennessee. Ill. Nat. Hist. Surv. Biol. Notes 113: 1–14.

Page, L.M. 1981a. The genera and subgenera of darters (Percidae: Etheostomatini). Occas. Pap. Mus. Nat. Hist. Univ. Kansas 90: 1–69.

Page, L.M. 1981b. Redescription of *Etheostoma australe* and a key for the indentification of Mexican *Etheostoma* (Percidae). Occas. Pap. Mus. Nat. Hist. Univ. Kansas 89: 1–10.

Page, L.M., M.E. Retzer & R.A. Stiles. 1982. Spawnin behavior in seven species of darters (Pisces: Percidae). Brimleyana 8: 135–143.

Page, L.M. & B.M. Burr. 1976. The life history of the slabrock darter, *Etheostoma smithi,* in Ferguson Creek, Kentucky. Ill. Nat. Hist. Surv. Biol. Notes 99: 1–12.

Page, L.M. & R.L. Mayden. 1981. The life history of the Tennessee snubnose darter, *Etheostoma simoterum,* in Brush Creek, Tennessee. Ill. Nat. Hist. Surv. Biol. Notes 117: 1–11.

Ramsey, J.S. & R.D. Suttkus. 1965. *Etheostoma ditrema,* a new darter of the subgenus *Oligocephalus* (Percidae) from springs of the Alabama River Basin in Alabama and Georgia. Tulane Stud. Zool. 12: 65–77.

Raney, E.C. & E.A. Lachner. 1943. Age and growth of johnny darters, *Boleosoma nigrum olmstedi* (Storer) and *Boleosoma longimanum* (Jordan). Amer. Midl. Nat. 29: 229–238.

Reeves, C.D. 1907. The breeding habits of the rainbow darter (*Etheostoma caeruleum* Storer), a study in sexual selection. Biol. Bull. (Woods Hole) 14: 35–59.

Scalet, C.G. 1971. Parasites of the orangebelly darter, *Etheostoma radiosum cyanorum* (Osteichthyes: Percidae). J. Parasitol. 57: 900.

Scalet, C.G. 1972. Food habits of the orangebelly darter, *Etheostoma radiosum cyanorum* (Osteichthyes: Percidae). Amer. Midl. Nat. 87: 515–522.

Scalet, C.G. 1973a. Stream movements and population density of the orangebelly darter, *Etheostoma radiosum cyanorum* (Osteichthyes: Percidae). Southwest. Nat. 17: 381–387.

Scalet, C.G. 1973b. Reproduction of the orangebelly darter, *Etheostoma radiosum cyanorum* (Osteichthyes: Percidae). Amer. Midl. Nat. 89: 156–165.

Scalet, C.G. 1974. Lack of piscine predation on the orangebelly darter, *Etheostoma radiosum cyanorum.* Amer. Midl. Nat. 92: 510–512.

Storer, D.H. 1845. Description of hitherto undescribed fishes. Proc. Boston Soc. Natur. Hist. 2: 47–49.

Strawn, K. 1955. A method of breeding and raising three Texas darters. Part I. Aquarium J. 26: 408–409, 411–412.

Strawn, K. 1956. A method of breeding and raising three Texas darters. Part II. Aquarium J. 27: 11–17, 31–32.

Turner, C.L. 1921. Food of the common Ohio darters. Ohio J. Sci. 22: 41–62.

Winn, H.E. 1957. Egg site selection by three species of darters (Pisces: Percidae). Anim. Behav. 5: 25–28.

Winn, H.E. 1958a. Observations on the reproductive habits of darters (Pisces: Percidae). Amer. Midl. Nat. 59: 190–212.

Winn, H.E, 1958b. Comparative reproductive behavior and ecology of fourteen species of darters (Pisces: Percidae). Ecol. Monog. 28: 155–191.

Received 31.8.1982 *Accepted 30.7.1983*

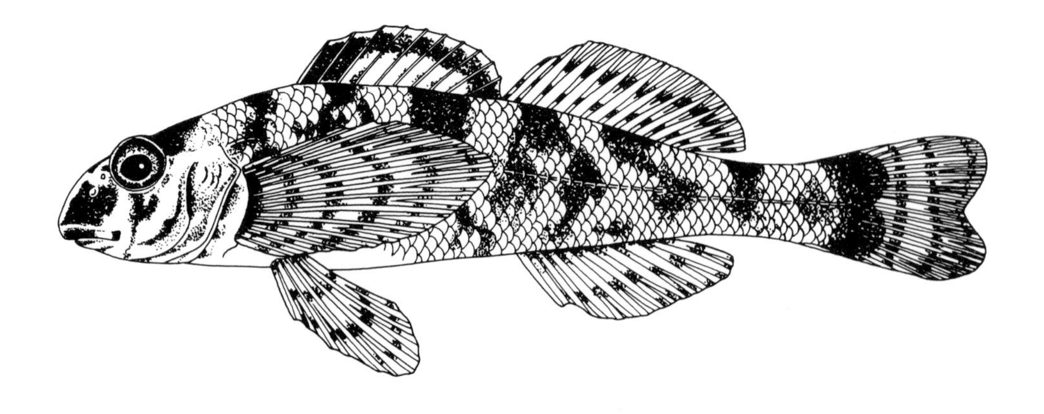

The harlequin darter *Etheostoma histrio*.

Life history of the bronze darter, *Percina palmaris*, in the Tallapoosa River, Alabama

Werner Wieland
Alabama Cooperative Fishery Research Unit, Auburn University, AL 36849, U.S.A.[1]

Keywords: Percidae, Ecology, Reproduction, Fecundity, Growth. Survival, Diet

Synopsis

Specimens of *Percina palmaris* from tributaries of the Tallapoosa River were examined for food, reproductive biology, age, and growth. Individuals were largely restricted to riffle areas with fast current and rubble-gravel substrate where fish density was higher than that reported for other species of *Percina*. Few other species were commonly taken with *P. palmaris* in its preferred habitat. Both sexes matured at one year of age and males appeared ready to spawn in mid February although females did not contain mature ova until March. Spawning occurred from early May to mid June and peaked at water temperatures of 18–20° C in late May. Although mature males were significantly larger than females, the oldest specimen examined was a female in its fourth year of life (3+). Generally *P. palmaris* is an opportunistic sight feeder although some selectivity in food items was suggested. Twenty-six food categories were identified from stomach contents over the course of the study. Diet consisted almost exclusively of immature insects and monthly diversity ranged from a minimum of 4 to a maximum of 20 insect families. *Percina palmaris* possesses some specialized as well as generalized life history attributes placing it near the middle of a generalized-specialized life history continuum for *Percina*. This is in agreement with its systematic status in this genus.

Introduction

The bronze darter, *Percina palmaris* (Bailey), and the gilt darter, *P. evides,* comprise the subgenus *Ericosma*. *Percina palmaris* is considered the more primitive form (Page 1974b) and is confined to the Tallapoosa and Coosa River systems of the Mobile Bay drainage. Crawford (1954) concluded that variation between populations of *P. palmaris* in the Coosa and Tallapoosa rivers was insufficient to warrant subspecific status. Denoncourt (1976) found significant differences between these populations but also did not suggest subspecific recognition. Denoncourt (1976) also found significant meristic differences between sexes, especially in the Tallapoosa River population.

Lee et al. (1980) listed 29 described species of *Percina,* and Page (1981) listed 30. Thorough life history accounts are available for only seven of these species. Little or no published life history information exists for 18 species and incomplete accounts are available for those remaining. No life history information has been published for *P. palmaris* despite the fact that it is not difficult to obtain specimens and the species was described over 40 years ago. In this paper are reported the diet, reproductive biology, age and growth of *P. palmaris* in relation to its congeners.

[1] Present address: Dept. of Biological Sciences, Mary Washington College, Fredericksburg, VA 22401, U.S.A.

David G. Lindquist & Lawrence M. Page (ed.), Environmental biology of darters. ISBN 90-6193-506-7

Methods and materials

Monthly seine collections of *P. palmaris* were made in Emuckfaw Creek, a tributary of the Tallapoosa River, Tallapoosa County, Alabama. Additional collections were made in Enitachopco and Hillabee creeks, Tallapoosa and Clay counties. Drift nets were placed in the stream one hour before each collection and removed before seining began. Surber samples (0.09 m^2) were taken prior to each collection in Emuckfaw and Enitachopco creeks. Drift and bottom samples were preserved in 5 percent formalin and returned to the laboratory where organisms were identified, counted, air dried and weighed. Fish collections were preserved in 10 percent formalin and later transferred to 50 percent isopropyl alcohol. Total length (mm TL), standard length (mm SL), weight (g) and sex were determined for each fish. All lengths reported here are SL unless otherwise noted. Scales were removed and age, scale radius and distance from scale center to annuli were determined. Growth comparisons are based on backcalculated lengths which were determined as described in Everhart et al. (1975). Survival was calculated by use of a Chapman & Robson estimator (Everhart et al. 1975).

A mark-recapture study was conducted on a small stock in the main channel of the Tallapoosa River. Individuals were marked by clipping the left pelvic fin. Five samples were made at one day intervals. Stock size was determined by both Schnabel and Schumacher estimates (Everhart et al. 1975). Confidence intervals were determined by treating the number of recaptures (R) as a Poisson variable (Ricker 1975, Appendix 2).

Ovaries were removed, air dried and weighed to the nearest 0.0001 g and the Gonadosomatic Index (GSI) was determined. Ovum diameter was measured to the nearest 0.01 mm with an ocular micrometer. Ova were then sorted by size and counted, and mean diameters for the twenty largest ova were determined. Developing ova were categorized according to size and color in a manner similar to that of Page & Smith (1970). The categories of ova were: less than 1.06 mm and white (considered to be developing); between 1.06 and 1.27 and pale yellow to yellow (considered immature); and larger than 1.27 mm and dark yellow to orange (considered mature). Percentage of differentiated ova (>0.55 mm) was based on actual counts. Terminology used on larval fish follows that of Hubbs (1943).

A length-weight relationship was determined for food organisms obtained from drift and bottom samples. In the case of those organisms which did not vary significantly in size an average weight was determined for each sampling period. Mean weights and length-weight relationships were then employed to reconstruct weights of stomach contents which had been identified, counted and measured for length.

Frequency of occurrence, numeric and gravimetric methods (Windell & Bowen 1978) were employed to analyze diet because most studies reported in the literature use one of these methods. These methods were deemed unsatisfactory in this study, for the following reasons. Frequency of occurrence is the percent of individuals with a particular food category in their stomachs. It is not, however, indicative of the relative importance of a particular food category in the diet. The numeric value, the percent of the total number of individual food items a particular food category comprises, gives a better indication of how important a particular food category might be in the diet relative to others; however, it fails to compensate for differences in the size of the food items. The gravimetric and volumetric methods correct for differences in size of the food item but also have their shortcomings. For example, large food items which occur in only one or two individuals will be interpreted as more important than smaller items which may appear in all individuals.

My conclusions about diet and food selection are primarily based upon an average of the weight percent method, a modification of the volume percent method of Larimore (1957). This method incorporates the gravimetric and frequency of occurrence methods and presents a more meaningful estimate of the importance of various food categories, especially when significant components of the diet do not occur in discrete units of uniform size, e.g., in *P. palmaris* an individual might contain items ranging in size from 0.023 mg (chironomids and simuliids) to 4.0 mg (siphlonurids and corydalids).

Two 24 h collections were made in which samples were taken every 3 h to determine daily feeding periodicity. Selectivity of feeding was examined using the Linear Food Selection Index (Strauss 1979). Similarity of diet between sexes was determined by an index of overlap (Schoener 1970) which Wallace (1981) decided was most representative.

Results and discussion

Habitat and associated species

In general, although exceptions did occur, a localized welldefined habitat for *P. palmaris* could be recognized from seine samples. Individuals were most common and largest in swift current (>30 cm sec^{-1}) at a depth of about 30 cm among cover, particularly river weed (*Podostemum*). Substrates consisted of rubble, gravel, and to a lesser extent, sand. When seining this habitat *P. palmaris,* was consistently taken alone. Movement of adults into pools during winter was not detected and although the largest individuals were commonly found in the swiftest current no apparent distinction could be made between habitat preferences of adults and juveniles. Numbers of individuals taken per unit effort was proportional to several factors such as area of the riffle sampled, size of substrate particles, flow volume and speed, and relative abundance of vegetation. Seining in other habitats, e.g., shallow riffles (<15 cm), runs at heads of riffles, shallow or deep runs with slow current (<30 cm sec^{-1}), spills below riffles, and pools, produced few or no specimens.

Species closely associated with *P. palmaris* were: *Notropis stilbius, N. callistius, Pylodictis olivaris, Noturus leptacanthus, Etheostoma jordani, Percina* (*Alvordius*) sp. and *Cottus carolinae.* Other associated species were *Campostoma anomalum, Hybopsis lineapunctata, Notropis chrysocephalus, N. gibbsi, N. venustus, Hypentelium etowanum, Micropterus salmoides, M. punctulatus, Etheostoma* (*Nanostoma*) sp. and *E. stigmaeum.* While *P. palmaris* usually was taken alone in its preferred habitat, when another species was present it usually was *Cottus carolinae.* All other species either occurred sporadically or only in pools below or at the head of riffles. From the large number of individuals taken *P. palmaris* seems to be quite successful in its preferred habitat.

Reproduction

Both sexes matured at one year of age, at which time the majority had reached at least 40 mm SL. Males yielded milt when squeezed as early as mid February (water temperature 8° C).

Females contained mature ova from mid March until July. One female was stripped of her eggs with strong pressure on 9 April (17° C) and by 6 May (18° C) only slight to moderate pressure was required to strip eggs. Monthly mean GSI and mean ova diameter are plotted in Figure 1.

Ovaluated ova ranged in size from 1.27 to 2.21 mm (mean = 1.75 mm), were orange in color, spherical, and contained an oil drop. No individuals examined in March contained ovulated ova. On May 6 one individual contained 52 ovulated ova. On 30 May two individuals contained 192 and 4 ovulated ova. On 19 June three indivuduals contained 10, 1 and 1 ovulated ova; however, in June, ovulated ova appeared somewhat deflated and ova in general appeared to be undergoing resorption. As evidenced by the presence of ovulated ova in females from 6 May to 19 June, the presence of eggs in the substrate and prolarvae from late June and July, spawning in *P. palmaris* occurs from early May to mid June. This period was characterized by a gradual rise in water temperature (7° C from 19 March to 19 June). A rapid rise in temperature might greatly shorten the spawning period as New (1966) found for *P. peltata* in New York. As indicated by GSI (Fig. 1), spawning peaked in May at 18–20° C and corresponded with maximum tubercular development of the male as reported by Denoncourt (1969). Time of spawning approximates that reported for other species of *Percina* (Petravicz 1938, Winn 1958a, 1958b, New 1966, Thomas 1970, Page & Smith 1970, 1971, Mathur 1973). However, members of the subgenus *Imostoma* have been

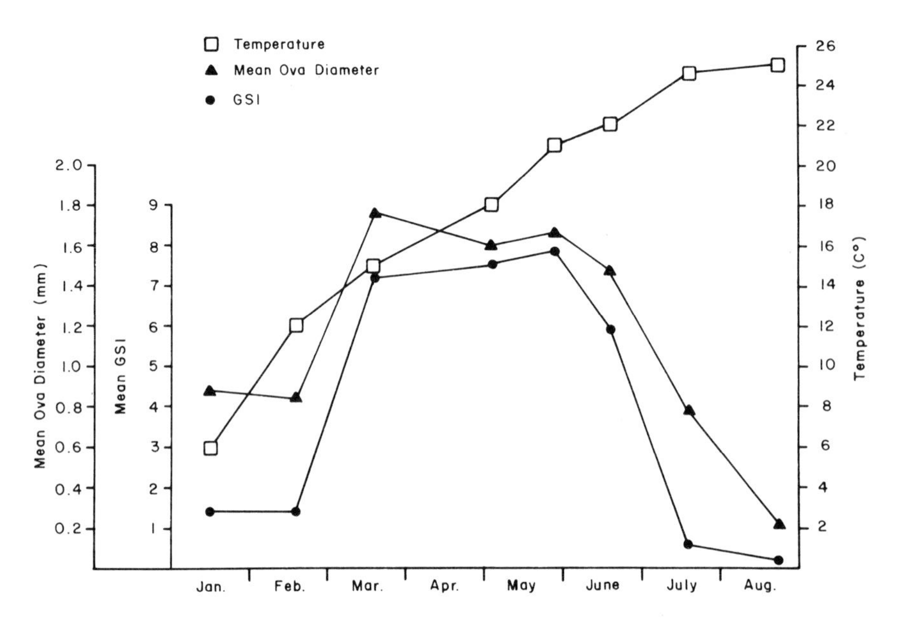

Fig. 1. Female reproductive cycle of *Percina palmaris* in Emuckfaw Creek.

reported to spawn earlier (Collette 1965, Starnes 1977).

Spawning was not observed in the field or in aquaria. On several occasions during the spawning season, individuals were captured and placed in aquaria (122 × 30 × 46 cm) where males were observed mounting females in the manner described by New (1966) for *P. peltata*. However, females appeared refractory, even though they were distended with eggs, and soon darted away. Care was taken to approximate the natural habitat, employing substrate from the stream from which fish were taken and maintaining a constant current. During these attempts to observe spawning, males established stationary territories which they defended.

Females collected in May and June (N = 35) contained from 37 to 404 mature ova and between 102 and 894 ova greater than 0.55 mm. The following formulae were determined as estimates of fecundity: $F = 107.86\ wt - 12.21$, $r = 0.41$; $F = 15.35\ SL - 547.65$, $r = 0.31$; $F = 12.92\ TL - 563.44$, $r = 0.32$; where F = number of mature ova; and r = correlation coefficient. All relationships were significant ($P<0.0001$). Equations to calculate fecundity as total number of differentiated ova (>0.55 mm) resulted in slightly higher correlation coefficients but, since it was evident that only those ova reaching the mature stage were spawned only those relationships are given.

Growth and demography

Larvae and eggs containing advanced embryos were taken from Emuckfaw and Enitachopco creeks on 19 and 26 June (water temperatures were about 21° C). Both were identified as those of *Percina* by myomere counts (Hogue et al. 1976) and information gained on pigmentation patterns from examination of percid larvae taken from the Tallapoosa and Coosa rivers by the Alabama Cooperative Fishery Research Unit. Due to differences in habitat preferrences and spawning times (Wieland unpublished) between *P. palmaris* and *P. (Alvordius)* sp., the only other species of *Percina* occurring in samples, these larvae and eggs were presumed to be those of *P. palmaris*. Eggs ranged in size from 1.7–1.9 mm in diameter. Embryos had a continuous median fin fold, a fully formed mouth, and the yolksac contained a single oil drop approximately 0.3 mm in diameter. Sand grains were adhering to several eggs. On 26 June a 4.1 mm TL prolarva was taken in Enitachopco Creek; pectoral fin buds were present and the yolksac was very small.

A 4.7 mm prolarva was taken from Emuckfaw Creek on 18 July (water temperature 24° C), and postlarvae 5.3–7.6 mm TL were obtained in bottom samples taken in riffles. In the largest of the postlarvae dorsal, anal and caudal fin folds and pectoral fin buds could be distinguished. No yolksac was evident and the gut appeared complete.

The largest postlarva, 8.5 mm TL, was taken on 28 August; water temperature was 25° C. In this specimen fin rays were visible and the gut contained food particles. No scales were present at 8.5 mm TL but squamation appeared complete in a 17 mm SL individual. From the relationship of scale radius to SL, scale formation is expected at about 9 mm.

The smallest *P. palmaris* taken by seining were obtained in mid July and estimated to be 4–6 weeks old (Fig. 2). Lengths of age group 1 and older are for backcalculated lengths. Like other species of *Percina* (e.g. Thomas 1970) *P. palmaris* exhibits rapid growth in its first year of life, achieving approximately 70 percent of its maximum length by the end of its first summer of growth. Males were significantly larger (P <0.01) than females at age 1 and 2 (Table 1). The oldest and largest individuals taken in this study were a female, age 3+ years at 67.4 mm and a male, age 2+ years at 70.3 mm, respectively. Page & Smith (1971) found that although their largest specimens of *P. phoxocephala* were males, sexual differences in mean SL were not significant. Thomas (1970) found no significant difference between sexes for *P. phoxocephala, P. shumardi, P. maculata* and *P. caprodes*. Page & Swofford (1984) report males larger than females as a characteristic of all territorial species of darters and attribute it to the selective advantage of being better able to hold territories and thus reproduce. Although having a faster rate of growth the lack of age 3 males in this study (Table 1) may indicate greater longevity for the slower growing females.

Conversion of TL to SL is given by the following: $TL = 1.195\ SL + 1.325$; $r = 0.99$; $P < 0.0001$ and $N = 440$. The overall length-weight relationship

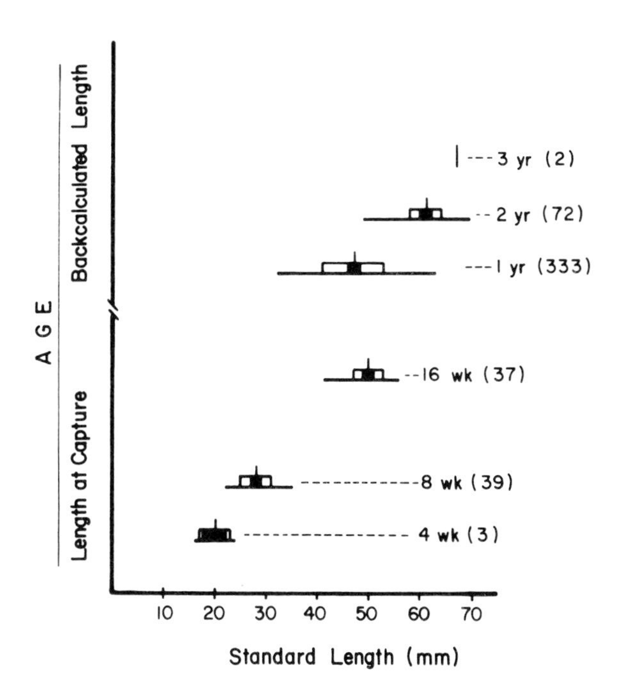

Fig. 2. Size distribution of *Percina palmaris*. Vertical line is mean, horizontal line is range, dark rectangle is ± 1 SE and light rectangle is ± 1SD. Smallest fish were taken in July (N).

Table 1. Differential growth between sexes of *Percina palmaris* in Emuckfaw and Enitachopco creeks based on backcalculated lengths.

Year class	Age					
	1		2		3	
	F	M	F	M	F	M
1976	47.2[a]	48.9	61.5	62.0	67.2	–
	1.7[b]	2.5*	3.8	4.3*	5.0	–
	(11)[c]	(18)	(11)	(18)	(2)	–
1977	44.3	51.0*	58.3	62.1*	–	–
	1.4	2.2*	3.1	4.1*	–	–
	(51)	(76)	(7)	(19)	–	–
1978	47.6	51.4*	56.2	60.6*	–	–
	1.7	2.2*	2.8	3.7*	–	–
	(75)	(24)	(10)	(7)	–	–
1979	42.4	45.7*	–	–	–	–
	1.1	1.5*	–	–	–	–
	(50)	(33)	–	–	–	–
Mean	45.3	49.7*	58.8	61.8*	67.2	–
	1.5	2.0*	3.3	4.1*	5.0	–
	(187)	(151)	(28)	(44)	(2)	–

* $p<0.01$.
[a] Mean standard length in mm.
[b] Mean weight in g.
[c] (N).

was determined as: Log^{10} Weight = (3.286 Log^{10} SL) − 5.298; R = 0.98; $P < 0.0001$ and N = 440. From growth data, the maximum SL predicted for *P. palmaris* by a Walford plot (Ricker 1975) would be approximately 73 mm. Denoncourt (1976) reported maximum lengths of 75 mm and 69 mm for male and female *P. palmaris* respectively. Thus the maximum SL for *P. palmaris* is less than the average of 92 mm for all species of *Percina* given by Page & Swofford (1984). Reduction of body size in darters is recognized as a derived characteristic (Page 1974b).

For pooled samples the overall sex ratio (female:male) was 1.3:1. However, the proportion of females to males varied with month, age, and collection site; in October it was 1:1.6, and in July and August it was 1.5:1, at age 0+, 1+, and 2+ the sex ratios were 1.6:1, 1.4:1, and 1:1.7, respectively (significant at $P < 0.05$).

Sufficient numbers of fish were obtained for the 1977 year class and in a single sample in 1980 to estimate survival. For the 1977 year class survival from age 1 through 3 was estimated at 17.8 percent (SE = 4.7) and for the 1980 sample containing age groups 1 to 3 (across year classes) survival was 19.6 percent (SE = 4.7).

An estimate of stock density was based on a multiple mark-recapture study on a small riffle area (approximately 125 m^2) in the main channel of the Tallapoosa River. The stock size was estimated at 130 (Schnabel method) to 133 (Schumacher method) or about 1 fish m^{-2} (95% C.I. = 1 fish 0.5–1.6 m^{-2}). Density estimates reported for other species of *Percina* are much lower, e.g., 1 fish 5.3 m^{-2} for *P. sciera* (Page & Smith 1970) and 1 fish 20.1–35.4 m^{-2} for *P. phoxocephala* (Thomas 1970; Page & Smith 1971).

Diet and feeding habits

Daily feeding periodicity data (Fig. 3) indicate *P. palmaris* is a diurnal sight feeder. However, the presence of food items in the stomach at 0200 h in October and July may indicate that some feeding continues after sunset. By 0500 h 90 percent of the stomachs observed on both dates were empty and of the remaining 10 percent it was evident from the condition of the stomach contents that nothing had been consumed recently. That *P. palmaris* is a sight feeder having a peak feeding period during daylight hours was further substantiated by observations in the field and in aquaria. Individuals were observed searching for food items which were taken from the substrate and within strands of *Podostemum*. Fish searched an area, stopping to scrutinize particular spots along the edges of stones and rubble and among patches of *Podostemum*. If something edible was observed the individual would quickly dart in and seize the item.

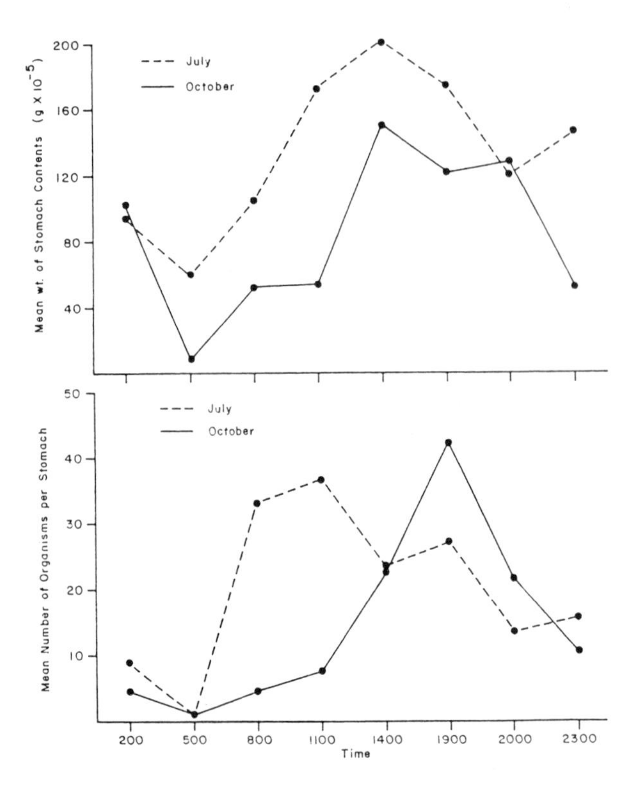

Fig. 3. Daily feeding periodicity in *P. palmaris*, expressed as mean number of organisms per stomach and as mean weight of stomach contents.

Mean number of organisms consumed peaked during the spawning period, as did the availability of food organisms (Fig. 4). A decrease in food intake occurred after spawning. Similar seasonal feeding peaks just before and during the spawning season were observed for *P. maculata* (Thomas 1970), *P. sciera* (Page & Smith 1970), *P. phoxocephala* (Page & Smith 1971), and several species of *Etheostoma* (Braasch & Smith 1967, Page 1974a,

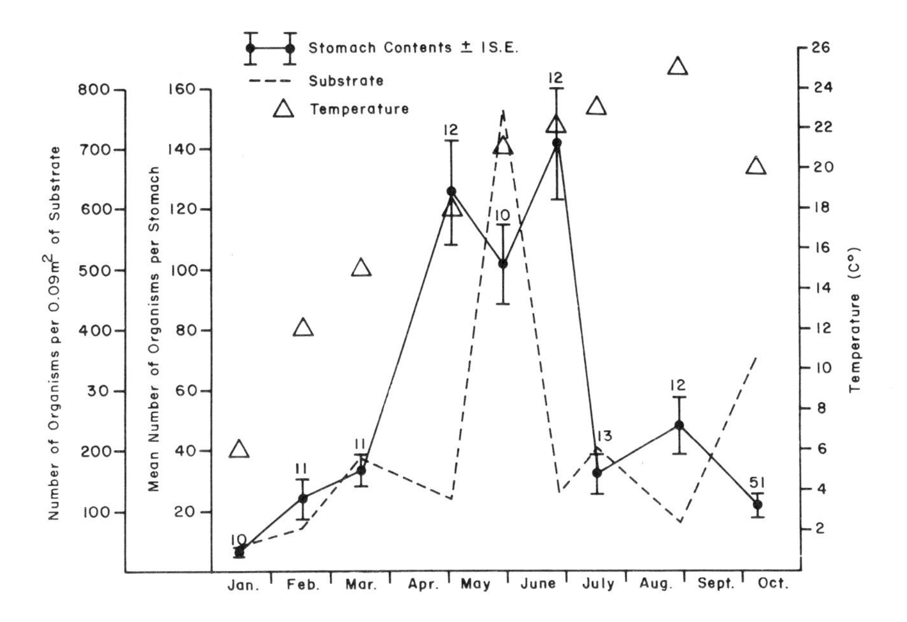

Fig. 4. Monthly mean number of organisms per stomach of *Percina palmaris,* and number of organisms 0.09 m^{-2} and water temperature in Emuckfaw Creek.

1975, Page & Burr 1976, Page & Mayden 1981). However, consumption by weight of stomach contents in *P. palmaris* was greatest in late summer after spawning (Fig. 5). The pattern of abundance (biomass) of organisms in the stream shown in Figure 5 closely follows the seasonal flucuations of insect biomass in streams as depicted by Hynes (1970). The feeding peaks observed based on numbers of organisms (Fig. 4) are the result of the greater availability of small early instars, neces-

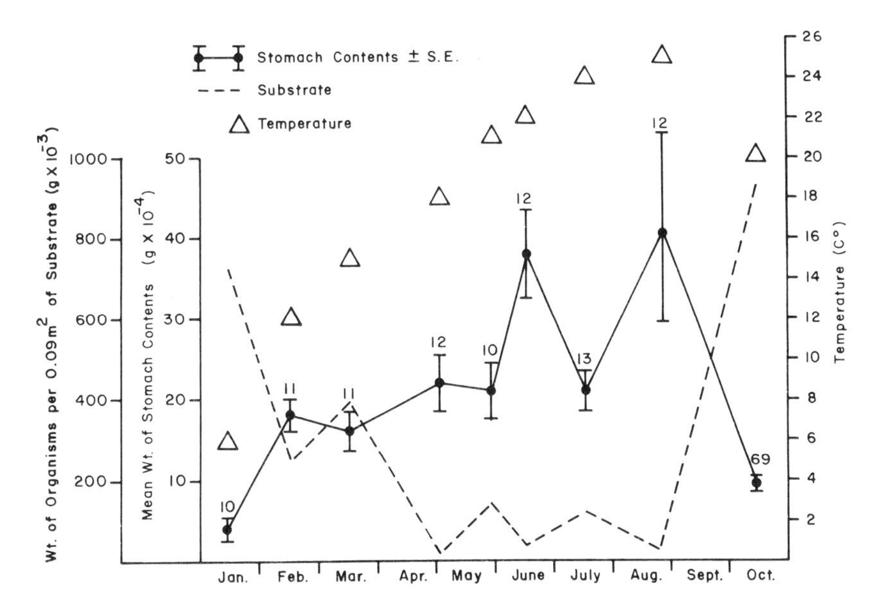

Fig. 5. Monthly mean weight of stomach contents of *Percina palmaris,* and weight of organisms 0.09 m^{-2} and water temperature in Emuckfaw Creek.

sitating a greater intake of numbers on the part of the fish to obtain what might be considered a maintenance diet. This demonstrates a problem with analyzing diets when only observing numbers of food items. Considering only the number of food items consumed can result in a misinterpretation of peak feeding. Additional evidence that increased consumption does not occur before or during spawning in several groups of fish can be found in Hoar (1969). Biomass is here considered a better measure of energy content than numbers of individuals and has been suggested as a reasonable substitute for calories gained from prey (Schoener 1974).

Increased consumption, based on intake of biomass (Fig. 5), occurs after spawning and corresponds with annulus formation in *P. palmaris.* Similar timing of annulus formation was reported in *P. peltata* by New (1966). This was a time of increasing water temperatures, and Mathur (1973) observed increased feeding with rise in water temperature for *P. nigrofasciata* in Halawakee Creek, Alabama. As temperature drops food intake (biomass) declines and remains low throughout the colder months even though biomass of stream organisms is highest at these times (Fig. 5).

Food categories indentified from stomachs were Hydracarina, Collembola, Baetidae, Heptageniidae, Siphlonuridae, Ephemerellidae, Aeshnidae, Perlidae, Elmidae, Psephenidae, Brachycentridae, Hydropsychidae, Leptoceridae, Polycentropodidae, Psychomyiidae, Rhyacophilidae, Corydalidae, Chironomidae, Simuliidae, Cyclorrhapha, Diptera pupae and adults, Hymenoptera, Decapoda, Gastropoda and fish eggs. Diet consisted almost exclusively of immature insects. The least diversity in diet was found in April 1978 and May 1979, when only four categories were present; the greatest diversity occurred in July 1980, when 20 of the above categories were present. Diet composition by month is summarized in Table 2 for food categories which occurred in greater than trace amounts for more than five out of the 12 months sampled. Comparison of diet with food availability by calculation of a Linear Food Selection Index (Strauss 1979) for major dietary components indicated random feeding in most instances (Table 3). Four insect families, Baetidae, Hydropsychidae, Chironomidae and Simuliidae, appeared in each monthly dietary sample. Baetids and chironomids were consistently eaten in proportion to their abundance. Negative values were often obtained for selection of hydropsychids. Three reasons can be given for these values; 1) inaccessibility of hydropsychids to *P. palmaris;* 2) presence of relatively large quantities of a particular organism, as was observed for hydropsychids, often results in low values of selection; 3) hydropsychids spend the majority of their time in their retreats (cases) and their movements are not easily detected by a sight feeder such as *P. palmaris.* Simuliids demonstrated consistently positive values of selection which were greatest from March through June; in May, a period corresponding with the time of spawning, *P. palmaris* fed almost exclusively on simuliids (Table 2). These high values, as a proportion of diet and as an index of selection, indicate a preference for or differential accessibility of simuliids. Simuliids are virtually restricted to riffles (Pennak 1953), as is *P. palmaris.* Prior to and during spawning male *P. palmaris* are establishing territories (or behaving agonistically if territories move with the individuals), attracting mates, and spawning, presumably with less time spent searching for food. Simuliids are a readily available and a highly conspicuous food item. Thus simuliids are not only heavily fed upon because they are readily available but by concentrating on them during spawning, *P. palmaris* is able to spend more time on spawning activities.

Similarity of diet between sexes was consistently high averaging 0.70 or greater depending on the method used for stomach analysis. Greatest overlap occurred during peak spawning (0.98–0.82) while lowest values were observed during the winter (0.38–0.69). Such similarity was expected because no difference in habitat preferences was detected between sexes.

Few apparent differences in diet between adults (>45 mm) and juveniles (<45 mm) were observed. Some differences were attributed to the greater mobility of adults and their ability to engulf a greater variety of prey sizes. The shift in size of prey items (small to large) as growth occurs is more

Table 2. Major monthly dietary components of *Percina palmaris* in Emuckfaw and Enitachopco creeks.

Food organism	20 MAR 1978	23 APR 1978	8 OCT 1978	15 JAN 1979	16 FEB 1979	19 MAR 1979	6 MAY 1979	30 MAY 1979	19 JUN 1979	18 JUL 1979	28 AUG 1979	30 JUL 1980
Ephemeroptera												
Baetidae	0.1[a]	47.6	47.8	2.6	0.4	0.2	0.4	7.1	14.1	16.7	30.6	7.7
	0.1[b]	58.6	34.4	1.8	0.3	0.2	0.4	6.8	14.5	15.4	33.3	8.1
	0.6[c]	31.5	26.1	4.9	1.1	0.3	0.3	5.8	11.6	22.8	51.1	15.6
	51[d]	80	81	30	18	9	25	80	100	100	100	68
Siphlonuridae	8.8	–	–	3.3	20.6	24.5	–	2.4	14.2	7.6	24.4	12.1
	14.6	–	–	5.2	22.3	26.5	–	4.3	10.7	9.1	12.5	18.7
	1.7	–	–	6.6	13.0	2.2	–	0.2	0.3	0.8	1.7	1.5
	16	–	–	30	64	54	–	10	42	31	50	24
Ephemerellidae	67.9	–	–	38.9	60.6	28.0	–	2.0	0.3	–	–	–
	64.8	–	–	73.3	58.9	23.7	–	3.2	0.2	–	–	–
	73.4	–	–	50.8	61.1	11.2	–	0.7	0.1	–	–	–
	84	–	–	60	100	91	–	20	8	–	–	–
Plecoptera												
Perlidae	1.8	–	4.0	19.6	8.7	–	–	–	–	–	1.6	0.1
	2.0	–	10.4	12.6	8.5	–	–	–	–	–	1.3	0.1
	2.3	–	0.4	16.4	7.0	–	–	–	–	–	0.3	0.2
	16	–	7	50	90	–	–	–	–	–	8	4
Coleoptera												
Elmidae	5.7	1.8	0.1	–	–	3.4	–	–	0.5	2.4	0.2	1.8
	4.8	4.5	0.1	–	–	2.3	–	–	0.4	2.3	0.2	1.9
	2.3	0.9	0.1	–	–	1.1	–	–	0.1	0.6	0.2	0.8
	16	20	1	–	–	27	–	–	17	23	8	13
Trichoptera												
Brachycentridae	–	–	1.3	2.6	0.2	0.6	–	–	0.1	0.7	–	1.9
	–	–	1.3	2.0	0.2	0.3	–	–	0.1	0.6	–	1.7
	–	–	0.3	1.6	0.4	0.5	–	–	0.1	0.4	–	1.7
	–	–	6	10	9	9	–	–	8	15	–	25
Hydropsychidae	10.8	–	4.0	20.0	7.6	7.1	0.4	0.5	10.3	38.4	17.9	32.8
	8.8	–	4.3	1.8	7.3	8.8	0.7	0.2	8.9	36.2	30.7	29.8
	4.6	–	2.2	3.3	4.1	2.7	0.1	0.1	1.5	28.0	3.8	24.8
	37	–	26	20	73	64	8	10	83	92	75	85
Diptera												
Chironomidae	4.5	31.2	5.1	10.7	0.4	10.3	0.7	1.0	6.0	1.8	1.7	2.8
	4.3	14.7	4.3	1.0	0.4	15.5	0.8	1.1	8.7	1.8	1.7	2.1
	12.7	24.1	15.9	4.9	2.6	19.9	1.0	1.5	9.6	5.4	5.2	7.4
	32	100	38	30	45	91	50	50	91	54	67	50
Simuliidae	0.2	19.4	23.4	2.0	1.2	24.7	98.5	86.4	54.0	10.9	11.9	15.9
	0.2	22.2	15.1	2.0	1.6	20.8	98.1	83.6	56.0	12.4	10.7	12.5
	1.2	43.5	52.6	9.8	10.0	61.7	97.7	90.2	76.6	37.4	36.3	39.7
	10	60	70	40	54	91	100	100	100	69	100	74
Number examined	19	5	70	10	11	11	12	10	12	13	12	67

[a] Average of the weight percent method.
[b] Gravimetric method.
[c] Numeric method.
[d] Frequency of occurrence method.

Table 3. Linear food selection index values for major dietary components of *Percina palmaris* in Emuckfaw Creek.

Food category	8 OCT 78	15 JAN 79	16 FEB 79	19 MAR 79	6 MAY 79	30 MAY 79	19 JUN 79	18 JUL 79	28 AUG 79	30 JUL 80
Ephemeroptera										
Baetidae	.38*[a]	.02	–	0	-.05	-.03	.13*	.14*	.30*	.06*
	.25*[b]	.01	–	0	-.05	-.03	.13*	.13*	.33*	.07*
	0[c]	.01	–	-.01	-.09	-.01	0	.18*	.48*	.12*
Siphlonuridae	–	-.07	.19*	–	–	-.01	-.44*	0	-.23	-.04
	–	-.05	.21*	–	–	.01	-.47*	.02	-.35*	.03
	–	-.30*	.08	–	–	-.01	-.07	-.03	-.07	-.06*
Trichoptera										
Brachycentridae	0	-.01	0	0	–	–	-.01	-.02	–	-.19*
	0	-.02	0	-.01	–	–	-.01	-.02	–	-.19*
	-.03*	-.02	-.03	-.02	–	–	-.06	-.09	–	-.20*
Hydropsychidae	.02*	-.51*	-.10	0	-.14	0	-.17	.18	.05	.05
	.02*	-.69*	-.11	.02	-.13	0	-.18	-.16	.18	.02
	-.14*	-.19	-.18	-.35*	-.06	0	-.13	-.02	-.07	-.02
Diptera										
Chironomidae	.04*	.10*	-.03	.09*	-.02	-.04	-.01	.01	0	.02*
	.03*	0	-.03	.14*	-.02	-.04	.02	.01	0	.01
	-.02	-.10	-.16	.02	-.10	-.13	-.30*	-.20	-.14	-.04
Simuliidae	.22*	.01	0	.24*	.69*	.26	.53*	.10*	.09	.12*
	.14*	.01	.01	.20*	.69*	.23	.55*	.12*	.07	.09*
	.31*	.06	.05	.46*	.47*	.13	.67*	.32*	.06	.27*

* Significantly different from 0 (P = 0.05).
[a] Average of the weight percent method.
[b] Gravimetric method.
[c] Numiric method.

conspicuous in larger species of darters such as *P. caprodes* (Thomas 1970).

Direct comparison of diet with other species of *Percina* was attempted. Allowing for differences in availability of various food items, which is not given in most studies, diet composition of *P. palmaris* is very similar to *P. maculata* (Thomas 1970), *P. phoxocephala* (Thomas 1970, Page & Smith 1971), *P. caprodes* (Thomas 1970), *P. shumardi* (Thomas 1970), *P. sciera* (Page & Smith 1970) and *P. nigrofasciata* (Mathur 1973). Immature insects predominate in the diets of all these species. Ephemeroptera, Chironomidae, Simuliidae and Trichoptera comprise the major dietary components. Simuliids consistently comprised a greater proportion of the diet of *P. palmaris* than reported for other species of *Percina*. Peak abundance of simuliids in the diet during spawning was not apparent for *P. phoxocephala* or *P. sciera* (Page & Smith 1970, 1971) as it was for *P. palmaris* (Table 2). Whether the greater abundance of simuliids in the diet of *P. palmaris* is due to abundance of simuliids in the streams sampled, their great abundance in the habitat of *P. palmaris*, or selection for this prey item is unknown. Simuliids, however, are widely abundant throughout North America (Pennak 1953, Merritt & Cummins 1978) and are presumed to have been common in the streams where these other studies were conducted.

Acknowledgements

I thank Richard K. Wallace for continued assistance in the field. Assistance in identification of larval fishes was provided by John S. Ramsey who also first suggested that I undertake this study. The Cooperative Fishery Research Unit is jointly sponsored by the Alabama Department of Conservation and Natural Resources, Auburn University, and the U.S. Fish and Wildlife Service.

References cited

Braasch, M.E. & P.W. Smith. 1967. The life history of the slough darter, *Etheostoma gracile* (Pisces, Percidae). Ill. Nat. Hist. Surv. Biol. Notes 58: 1–12.

Collette, B.B. 1965. Systematic significance of breeding tubercles in fishes of the family Percidae. Proc. U.S. Nat. Mus. 117: 567–614.

Crawford, R.W. 1954. Status of the bronze darter, *Hadropterus palmaris*. Copeia 1954: 235–236.

Denoncourt, R.F. 1969. A systematic study of the gilt darter *Percina evides* (Jordan and Copeland) (Pisces, Percidae). Ph. D. Thesis, Cornell University, Ithaca, 216 pp.

Denoncourt, R.F. 1976. Sexual dimorphism and geographic variation in the bronze darter, *Percina palmaris* (Pisces: Percidae). Copeia 1976: 54–59.

Everhart, W.H., A.W. Eipper & W.D. Youngs. 1975. Principles of fishery science. Cornell University Press, Ithaca. 288 pp.

Hoar, W.S. 1969. Reproduction. pp. 1–72. *In*: W.S. Hoar & D.J. Randall (ed.) Fish Physiology, Vol II, Reproduction and Growth, Bioluminescence, Pigments and Poisons, Academic Press, New York.

Hogue, J.J., Jr., R. Wallus & L.K. Kay. 1976. Prelimiary guide to the identification of larval fishes in the Tennessee River. Tech. Note B19. TVA, Division of Forestry, Fisheries, and Wildlife Development, Norris.

Hubbs, C.L. 1943. Terminology of early stages of fishes. Copeia 1943: 260.

Hynes, H.B.N. 1970. The ecology of running waters. University of Toronto Press, Toronto. 555 pp.

Larimore, W.B. 1957. Ecological life history of the warmouth (Centrarchidae). Ill. Nat. Hist. Surv. Bull. 27: 1–83.

Lee, D.S., C.R. Gilbert, C.H. Hocutt, R.E. Jenkins, D.E. McAllister & J.R. Stauffer, Jr. 1980. Atlas of North American freshwater fishes. N.C. State Mus. Nat. Hist., Raleigh. 867 pp.

Mathur, D. 1973. Food habita and feeding chronology of the blackbanded darter, *Percina nigrofasciata* (Agassix), in Halawakee Creek, Alabama. Trans. Amer. Fish. Soc. 102: 48–55.

Merritt, R.W. & K.W. Cummins. 1978. An introduction to aquatic insects of North America. Kendall Hunt, Dubuque. 441 pp.

New, J.G. 1966. Reproductive behavior of the shield darter, *Percina peltata peltata,* in New York. Copeia 1966: 20–28.

Page, L.M. 1974a. The life history of the spottail darter, *Etheostoma squamiceps,* in Big Creek, Illinois, and Ferguson Creek, Kentucky. Ill. Nat. Hist. Surv. Biol. Notes 89: 1–20.

Page, L.M. 1974b. The subgenera of *Percina* (Percidae: Etheostomatini). Copeia 1974: 66–86.

Page, L.M. 1975. The life history of the stripetail darter, *Etheostoma kennicotti,* in Big Creek, Illinois. Ill. Nat. Hist. Surv. Biol. Notes 93: 1–15.

Page, L.M. 1981. The genera and subgenera of darters (Percidae, Etheostomatini). Univ. Kansas Mus. Nat. Hist. Occ. Pap. 90: 1–69.

Page, L.M. & B.M. Burr. 1976. The life history of the slabrock darter, *Etheostoma smithi,* in Ferguson Creek, Kentucky. Ill. Nat. Hist. Surv. Biol. Notes 99: 1–12.

Page, L.M. & R.L. Mayden. 1981. The life history of the Tennessee subnose darter, *Etheostoma simoterum,* in Brush Creek, Tennessee. Ill. Nat. Hist. Surv. Biol. Notes 117: 1–11.

Page, L.M. & P.W. Smith. 1970. The life history of the dusky darter, *Percina sciera,* in the Embarras River, Illinois. Ill. Nat. Hist. Surv. Biol. Notes 69: 1–15.

Page, L.M. & P.W. Smith. 1971. The life history of the slenderhead darter, *Percina phoxocephala,* in the Embarras River, Illinois. Ill. Nat. Hist. Surv. Biol. Notes 74: 1–14.

Page, L.M. & D.L. Swofford. 1984. Evolution in darters: morphological correlates of ecological specializations. Env. Biol. Fish. 11: 139–159.

Pennak, R.W. 1953. Fresh-water invertebrates of the United States. Ronald Press, New York. 769 pp.

Petravicz, W.P. 1938. The breeding habits of the black-sided darter, *Hadropterus maculatus* Girard. Copeia 1938: 40–44.

Ricker, W.E. 1975. Computation and interpretation of biological statistics of fish populations. Fish. Res. Board Can. Bull. 191: 1–382.

Schoener, T.W. 1970. Non-synchronous spatial overlap of lizards in patchy habitats. Ecology 51: 408–418.

Schoener, T.W. 1974. Some methods for calculating competition coefficients from resource-utilization spectra. Amer. Nat. 108: 332–340.

Starnes, W.B. 1977. The ecology and life history of the endangered snail darter, *Percina* (*Imostoma*) *tanasi* Etnier. Tech. Rep. Tenn. Wildl. Res. Agency: 77–52.

Strauss, R.E. 1979. Reliability estimates for Ivlev's Electivity Index, the Forage Ratio, and a proposed Linear Index of Food Selection. Trans. Amer. Fish. Soc. 108: 344–352.

Thomas, D.L. 1970. An ecological study of four darters of the genus *Percina* (Percidae) in the Kaskaskia River, Illinois. Ill. Nat. Hist. Surv. Biol. Notes 70: 1–18.

Wallace, R.K. 1981. An assessment of diet-overlap indexes. Trans. Amer. Fish. Soc. 110: 72–76.

Windell, J.T. & S.H. Bowen. 1978. Methods for study of diets based on analysis of stomach contents. pp. 219–226. *In*: T.

Bagenal (ed.) Methods for the Assessment or Fish Production in Fresh Waters, IBP Handbook No. 3, Blackwell Scientific Publ., London.
Winn, H.E. 1958a. Comparative reproductive behavior and ecology of fourteen species of darters (Pisces-Percidae). Ecol. Monog. 20: 155–191.
Winn, H.E. 1958b. Observations on the reproductive habits of darters (Pisces-Percidae). Amer. Midl. Natur. 59: 190–212.

Received 20.8.1982 *Accepted 28.7.1983*

Temperature selection and critical thermal maxima of the fantail darter, *Etheostoma flabellare*, and johnny darter, *E. nigrum*, related to habitat and season

Christopher G. Ingersoll[1] & Dennis L. Claussen
Department of Zoology, Miami University, Oxford, OH 45056, U.S.A.

Keywords: Temperature preference, Temperature tolerance, Seasonal effects, CTMax, Fishes, Etheostomatini

Synopsis

Riffle dwelling fantail darters (*Etheostoma flabellare*) selected lower temperatures in winter (19.3°C) compared to pool dwelling johnny darters (*E. nigrum*; 22.0°C). A similar trend was evident in summer tests (fantail darters, 20.3°C; johnny darters, 22.9°C). Summer tested animals selected higher temperatures than winter tested animals maintained at the same acclimation temperature and photoperiod. When tested together in the same gradient, both species appeared not to thermoregulate, but tended to avoid each other. Critical thermal maxima (CTMax) did not differ between seasons for either species (fantail darters, 31.1°C winter, 31.3°C summer; johnny darters, 30.9°C winter, 30.5°C summer). Differences in the thermal responses of these darters correlated with differences in their respective habitats.

Introduction

Ectothermic animals, when presented a choice of temperatures may behaviorally thermoregulate by selecting a relatively narrow range of temperatures. This range, as determined by the frequency distribution within a thermal gradient, is referred to as a thermal preferendum (Fry 1947). Investigators have determined preferred temperatures of laboratory animals by two methods; the acute thermal preferendum (obtained within two hours or less after animals have been placed in a gradient), and the final temperature preferendum (obtained after 24 to 96 h in a gradient) (Reynolds & Casterlin 1979). For recent reviews of the literature see: Precht et al. (1973), Coutant (1977), Reynolds (1977), Spotila et al. (1979), Hutchison & Maness (1979) and Cherry & Cairns (1982).

The present study was initiated to compare the acute temperature preferences of two species of darters occupying different habitats within a small first order stream in southwestern Ohio. Fantail darters (*Etheostoma flabellare*) inhabit riffles, whereas johnny darters (*E. nigrum*) are found in adjacent pools. Critical thermal maxima (the temperature at which coordinated locomotor movements are lost) were also determined to further elucidate the thermal biology of these fishes.

[1] Present address: Department of Zoology and Physiology, University of Wyoming, Laramie, WY 82071, U.S.A.

David G. Lindquist & Lawrence M. Page (ed.), Environmental biology of darters. ISBN 90-6193-506-7

Materials and methods

Collection and maintenance of animals

Fish were collected 24 February, 1982 (temperature 2.0° C, oxygen 10.8 mg O_2 l^{-1}, pH 8.2), and 15 June, 1982 (temperature 15.8° C, oxygen 8.65 mg O_2 l^{-1}, pH 8.4), by seining Harker's Run, a tributary of Four Mile Creek in Butler County, Ohio. The stream also contained greenside (*E. blennioides*), banded (*E. zonale*), rainbow (*E. caeruleum*) and orangethroat (*E. spectabile*) darters. Other common species included the silverjaw minnow (*Ericymba buccata*), common stoneroller (*Campostoma anomalum*), creek chub (*Semotilus atromaculatus*), blacknose dace (*Rhinichthys atratulus*) and northern hog sucker (*Hypentelium nigricans*).

Twenty fantail and johnny darters from each collection were acclimated to 15.0 ± 1.0° C for a minimum of two and a maximum of three weeks in 23 liter Nalgene tanks. Fish were kept on a 12:12 photoperiod and fed commercial fish food ad libitum. Water was renewed every other day with aged (dechlorinated) tap water and aerated continuously.

Critical thermal maxima (CTMax)

Critical thermal maxima were determined between 0800 and 1200 hours. Individual fish were placed in 2500 ml glass beakers containing 1250 ml of water at 15.0° C. Water in the testing apparatus was aerated continuously and heated at a rate of 1° C min^{-1} with a heating coil surrounding the beaker and controlled by a rheostat. As water temperature approached the CTMax, the fish would begin to lose muscular coordination and lay motionless at the bottom of the beaker. They would then resume swimming and begin violent spasms. The water temperature at which the onset of spasms occurred was taken to be the CTMax (Kowalski et al. 1978). Differences among species and season were analyzed with two way ANOVA with mean separation by Duncan's New Multiple Range Test (Helwig & Council 1979).

Temperature selection

The fish were tested in a copper tank (88.9 cm × 15.9 cm × 6.35 cm) lined with Saran Wrap and filled to a depth of 2 cm with water. Each end of the copper tank was immersed in a plastic chamber. Water in one plastic chamber was heated and pumped through a copper tube soldered to the underside of the copper tank. This water was then passed through a copper coil immersed in the second plastic tank, which contained a refrigerated antifreeze solution that was cooled with a Frigid Midget Model CE 11 liquid cooling unit. The water was finally returned to the hot chamber via another copper tube running the length of the copper tank. These two copper tubes were arranged so that the flow of water within was counter current. Temperatures were measured by placing seven thermocouples through holes in a plastic cover which was marked off in 17 five-cm long zones.

All tests were conducted between 0800 and 1200 h. Six fish of one species were placed in the copper tank with no gradient (as controls) and, after 15 min, the position of each fish in the tank was recorded every minute for 30 min. These fish were removed while the gradient was established, then returned to the tank, and, after 15 min, the position of each fish in the thermal gradient was again recorded every minute for 30 min. Each test was repeated with a second set of six fish. Observers tried to disturb the fish as little as possible during and between readings. No attempt was made to distinguish among individuals when recording data and if mortality occurred, those data were discarded. Temperature preferenda of fantail and johnny darters were also determined by placing three individuals of each species together in the gradient and observing their responses as described above.

Distributions of fish in the tank were analyzed with the Chi-square goodness of fit test. Inasmuch as the established gradient was not perfectly linear, the distribution of fish expected in a particular range of temperatures was adjusted to the distance of that temperature range within the tank. For example, in Figure 1 the distance in the gradient between 20.0–22.9° C was approximately three

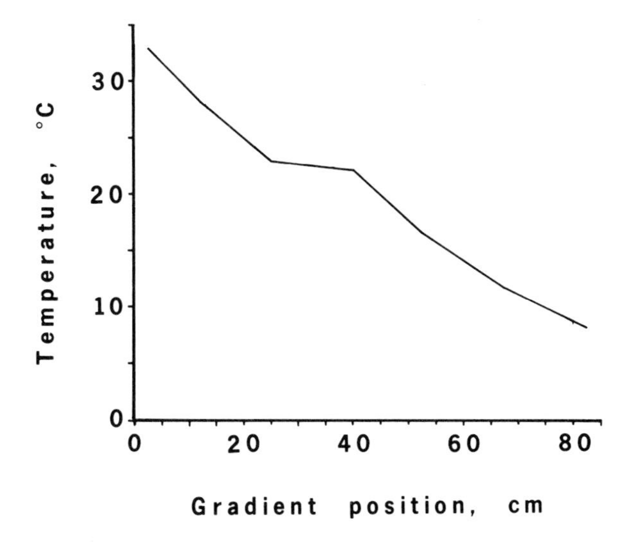

Fig. 1. The thermal gradient for winter tested *E. nigrum*. This gradient is representative of the several gradients established during the course of this study. Gradient position is here described as a distance from the 'hot' end.

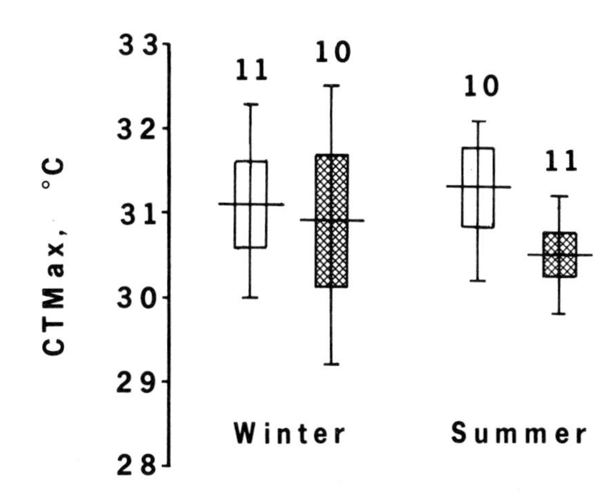

Fig. 2. The critical thermal maxima of *E. flabellare* (open rectangles) and *E. nigrum* (cross-hatched rectangles) in relation to season. The horizontal lines indicate the means, the vertical lines the ranges, and the rectangles the 95% confidence limits about the means. Sample size is given above each symbol.

times greater than the distance between 17.0–19.9° C, so the number of fish expected for random distribution between 20.0–22.9° C would be three times greater than the number of fish expected between 17.0 and 19.9° C. Gradient temperatures ranged from 7–33° C in winter and 10–29° C in summer testing.

Results

Critical thermal maxima

The CTMax of winter tested johnny darters (30.9° C) and fantail darters (31.1° C) were similar; however, the CTMax of summer tested johnny darters was 30.5° C compared to 31.3° C for fantail darters (Fig. 2). There was no significant difference in CTMax by season ($p = 0.586$), but johnny darters had a significantly lower CTMax compared to fantail darters ($p = 0.035$) with no interaction found between season and species ($p = 0.261$) as determined by two way ANOVA. CTMax was not correlated with body length.

Temperature selection of species tested separately

Both species tended to congregate at the ends of the tank in the absence of a thermal gradient (see Fig. 3). This might have been the result of fish attempting to seek cover. All such resulting distributions were significantly non-random (χ^2; $p < 0.001$). In contrast, darters in the established thermal gradient selected the middle of the tank, and again had a distribution which was significantly nonrandom (χ^2; $p < 0.001$, Fig. 4). During the first

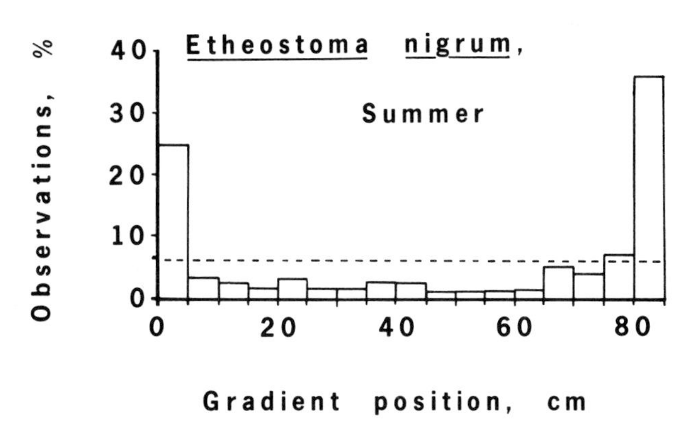

Fig. 3. Darter (*E. nigrum*, 12 fish, summer) position within the tank in the absence of a thermal gradient. The dashed line indicates the expected random distribution. Gradient position is described as a distance from the left end of the tank, which becomes the 'hot' end once a gradient is established.

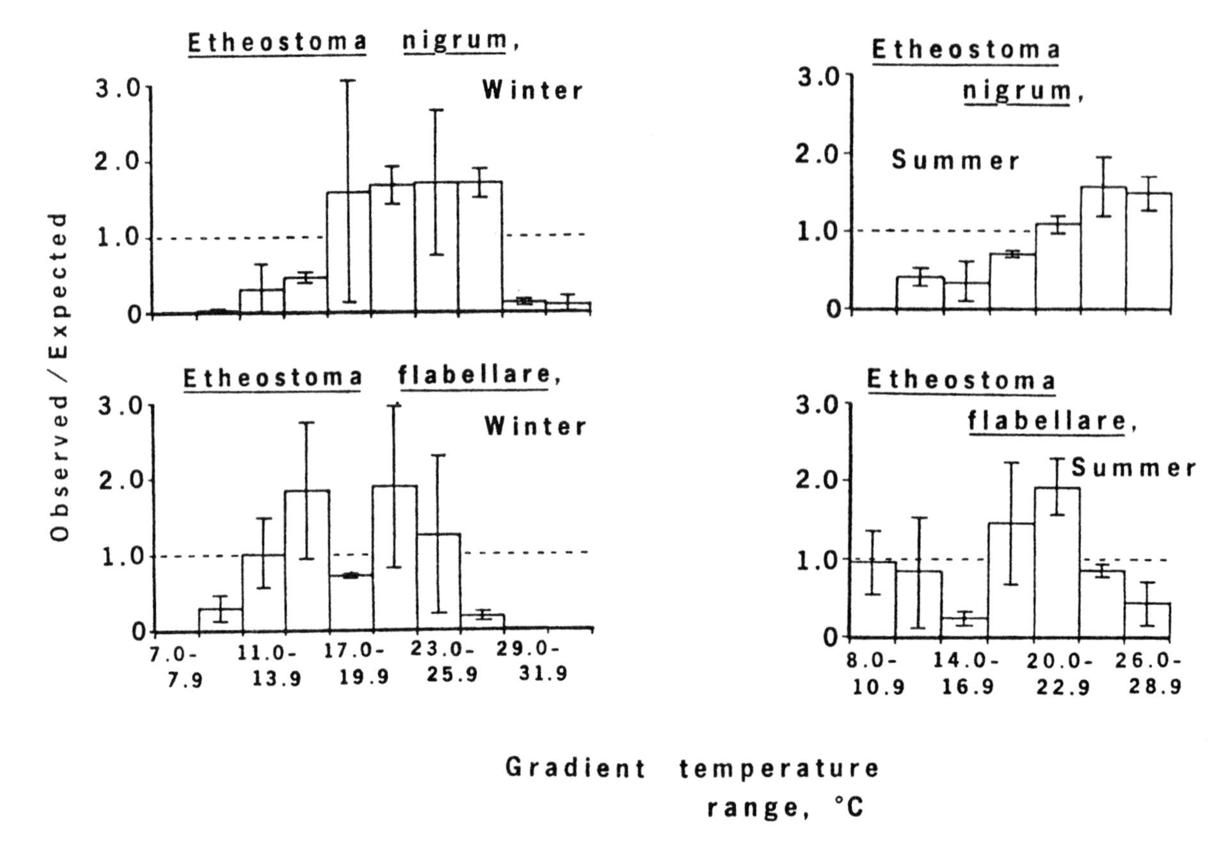

Fig. 4. Observed/expected ratios for winter and summer tested *E. nigrum* (10 fish) and *E. flabellare* (12 fish) in the established thermal gradient. Vertical lines connect the values from the two replications and the histograms depict the mean of these values for each test. The dashed lines indicate the expected random distributions. Gradient temperatures ranged from 7–33° C in winter and 10–29° C in summer testing.

15 min after the fish were introduced into the gradient, they would first explore the extremes of the temperature gradient before settling down. Observations during the next 30 min were used to determine the acute thermal preference of these species. The darters seldom selected a single temperature within the gradient, but rather tended to move between higher and lower temperatures with the preferred temperature being any of several measures of central tendency. Temperatures selected by johnny darters showed a unimodal distribution, whereas the patterns for fantail darters were less regular and bimodal (Fig. 4).

The temperatures selected by johnny darters were more than 2.5° C higher than the respective values for fantail darters in both winter and summer (Table 1). These temperatures selected in the gradient by johnny darters were significantly different from those selected by fantail darters in both winter ($\chi^2 = 99.34$, df = 4, $p < 0.001$; Fig. 4) and summer ($\chi^2 = 67.79$, df = 4, $p < 0.001$; Fig. 4). The median temperature and most frequently occupied temperature as indicated by the 80% modal range of observed temperatures selected, were also higher for johnny darters during both seasons (Table 1). Temperatures selected by each species increased approximately 1° C from winter to summer testing (Table 1, Fig. 4).

Temperature selection of species tested together

Distributions of control fish in the absence of a gradient were not random (χ^2; $p < 0.001$); each species tended to avoid the other by staying at opposite ends of the tank (Fig. 5). Temperatures occupied by johnny darters in the gradient

Table 1. Mean (standard error), median and 80% modal range of the observed temperatures selected by *E. flabellare* (12 fish) and *E. nigrum* (10 fish) tested separately.

Species	Mean (°C)	Median (°C)	80% modal range (°C)
E. flabellare, winter	19.3 (0.222)	21.8	12.2–23.2
E. nigrum, winter	22.0 (0.214)	22.7	17.6–26.8
E. flabellare, summer	20.3 (0.199)	20.9	16.2–24.5
E. nigrum, summer	22.9 (0.189)	23.0	18.9–28.2

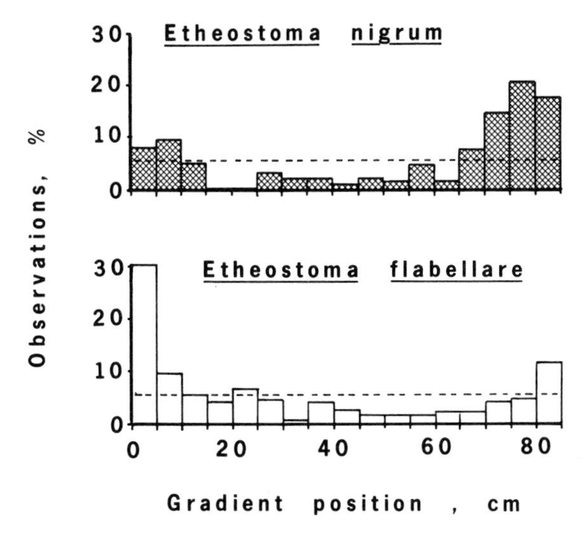

Fig. 5. Position of *E. nigrum* (6 fish, cross hatched histogram) and *E. flabellare* (6 fish, open histogram) tested together in the absence of a thermal gradient. The dashed lines indicate the expected random distributions.

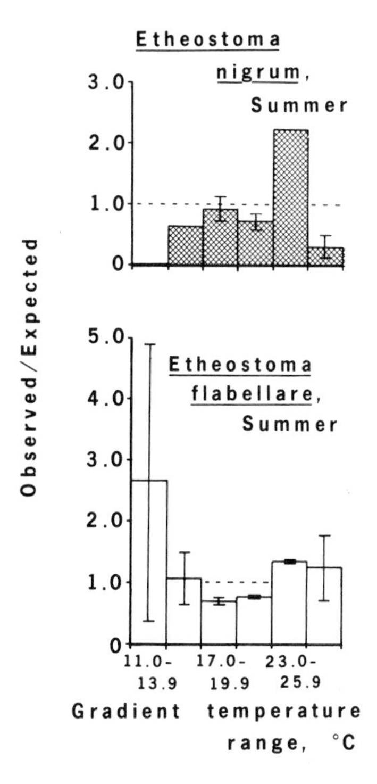

Fig. 6. Observed/expected ratios for summer tested *E. nigrum* (6 fish, cross hatched histogram) and *E. flabellare* (6 fish, open histogram) tested together in the established thermal gradient. Vertical lines connect the values from the two replications and the histograms depict the mean of these values for each test. The dashed lines indicate the expected random distribution. The ratios for the two replications were identical in two cases for *E. nigrum*. Gradient temperatures ranged from 11° C to 29° C.

were significantly different ($\chi^2 = 44.03$, df = 4, $p < 0.001$; Fig. 6) from those of fantail darters. Fantail darters did not select a single range of temperatures but rather occupied the extremes of the gradient, apparently to avoid the johnny darters which occupied temperatures in the middle of the temperature range. Johnny darters avoided temperatures of 26.0–28.9° C that were previously selected when tested separately (Fig. 4, 6). The standard error of the mean and 80% modal range of observed temperatures selected decreased for johnny darters and increased for fantail darters tested together (Table 2). These data suggest that the two darter species were not thermoregulating, but merely avoiding each other.

Discussion

Hill & Matthews (1980) first successfully demonstrated temperature selection in darters, using a cylindrical apparatus. They suggested that the lack of temperature selection by darters in earlier stud-

Table 2. Mean (standard error), median and 80% modal range of the observed temperatures selected by *E. flabellare* (6 fish) and *E. nigrum* (6 fish) tested together.

Species	Mean (°C)	Median (°C)	80% modal range (°C)
E. flabellare, summer	21.1 (0.310)	22.4	12.8–28.0
E. nigrum, summer	21.6 (0.198)	22.8	14.8–28.0

ies (Cherry et al. 1975, Stauffer et al. 1976) may have been the result of the fish attempting to hide in the corners of the rectangular tanks. However, the shape of the tank cannot explain the lack of temperature selection in the studies by Cherry et al. (1975) and Stauffer et al. (1976) since our darters tested in a rectangular tank did clearly select specific temperature ranges.

The tendency of the darters to occupy the ends of the chamber in the absence of a thermal gradient complicated the statistical analysis. We adopted a conservative approach and accepted distributional data as evidence of temperature selection only if the resulting χ^2 values were significantly different from both the no-gradient control and a random distribution. The data here reported (χ^2 values and Fig. 4, 6) are based on an 'expected' random distribution.

Temperature preference and tolerance are regulated by complex interactions of photoperiod, temperature, diel rhythms and annual rhythms (Hutchison & Maness 1979). Both species of darters selected warmer temperatures during summer tests even though winter and summer animals were acclimated to the same photoperiod and acclimation temperature. This suggests that temperature preference in these species may follow an endogenous circannual rhythm. The range of temperature in the gradient was greater during winter testing; however, the difference was probably not responsible for the seasonal shift in thermal preference, since winter tested animals avoided temperatures greater than 28.9° C and less than 11.0° C. Inherent seasonal rhythms in temperature preference or thermal tolerance have been demonstrated for some species of fish (Sullivan & Fisher 1953, Hoar 1955, Tyler 1966, Kowalski et al. 1978). Nevertheless, other investigators have found no significant inherent seasonal shift in the thermal preference of other fish species (McCauley & Tait 1970, Garside & Tait 1958, DeVlaming 1971, McCauley & Huggins 1979).

Our CTMax values for johnny and fantail darters are in close agreement with values of 32.1° C and 30.7–31.4° C respectively reported for these darters by Kowalski et al. (1978). These authors suggested that the lower CTMax of the johnny darters may reflect a difference in habitat from the riffle dwelling fantail darter. A seasonal increase in the CTMax of johnny darters demonstrated by these authors, however, was not evident in the present study.

Thermoregulation by ectotherms involves exploitation of a thermally heterogenous environment (Reynolds 1979). We observed temperatures in Harker's Run to vary up to 6° C diurnally during the summer months and, on occasion, pools were 3° C warmer than adjacent riffles, although riffles and pools were often isothermal. Small streams in the summer have been reported to fluctuate up to 6° C diurnally with shallow water exhibiting more of a daily variation in temperature (Hynes 1970). Neel (1951) observed slight differences between temperatures of riffles and pools, with riffles dropping to lower temperatures at night. Mean annual temperatures of pools in a small stream located in southwestern Ohio were reported to be warmer than adjacent riffles (Lynn 1972). The warmer temperatures selected by johnny darters may reflect its field distribution in pools, whereas the cooler and more variable temperatures selected by fantail darters correspond to the distribution of this species in the thermally more variable riffle habitat. The higher CTMax of the fantail darters may also reflect the rapid temperature fluctuations of riffles. Matthews & Styron (1981) report similar findings with oxygen tolerance of fantail darters. Fish inhabiting intermittent headwaters of a stream were more tolerant of low oxygen stress than fish from the stable mainstream. Hill & Matthews (1980) attributed differences in thermoselection by two species of darters to the thermal stability of their respective habitats. They found orangethroat darters (*E. spectabile*) collected from a thermally stable spring to show less variation in temperature selection compared to the orangebelly darters (*E. radiosum*) collected from a thermally more labile habitat.

Reynolds & Casterlin (1979), Magnuson et al. (1979) and Mathur et al. (1982) suggest that fish may partition thermal resources to reduce competitive interactions. Gehlbach et al. (1978) reported that temperatures selected by the Comanche Springs pupfish (*Cyprinodon elegans*) are lower when they are tested sympatrically with Pecos gambusia (*Gambusia nobilis*). Barans & Tubb (1973) reported similar findings for two species of *Notropis*. Johnny and fantail darters will defend a territory from other darters even during non-reproductive periods (Winn 1958) and the lack of temperature selection by these species tested together may reflect this antagonistic behavior rather than a shift in thermal niche. Temperature selection by our darters may also have been influenced by intra-specific interactions. One winter tested fantail darter was observed to nip the side of a second individual with the attacked fish swimming 10 cm into the warmer portion of the gradient. Social interactions have also been observed with the bluegill (*Lepomis macrochirus*) tested in a thermal gradient where subordinant individuals selected nonpreferred temperatures in the presence of larger, dominant fish (Beitinger & Magnuson 1975).

Laboratory temperature selection by johnny and fantail darters appears to be influenced by season, inter-specific and intra-specific interactions. Nevertheless, both the CTMax and the thermal preferenda of these fishes correlate well with their respective pool and riffle habitats.

Acknowledgments

We would like to express our appreciation to Michael J. Mac and two anonymous reviewers for their valuable suggestions.

References cited

Barans, C.A. & R.A. Tubb. 1973. Temperatures selected seasonally by four fishes from western Lake Erie. J. Fish. Res. Board Can. 30: 1697–1703.

Beitinger, T.L. & J.J. Magnuson. 1975. Influence of social rank and size on thermoselection behavior of bluegill (*Lepomis macrochirus*). J. Fish. Res. Board Can. 32: 2133–2136.

Cherry, D.S., K.L. Dickson & J. Cairns, Jr. 1975. Temperature selected and avoided by fish at various acclimation temperatures. J. Fish. Res. Board Can. 32: 485–491.

Cherry, D.S. & J. Cairns, Jr. 1982. Biological monitoring. Part V–Preference and avoidance studies. Water Res. 16: 263–301.

Coutant, C.C. 1977. Compilation of temperature preference data. J. Fish. Res. Board Can. 34: 739–745.

DeVlaming, V.L. 1971. Thermal selection behavior in the estuarine goby *Gillichthys mirabilis* Cooper. J. Fish Biol. 3: 277–286.

Fry, F.E.J. 1947. Effects of the environment on animal activity. Ontario Fish. Res. Lab. 68: 1–62.

Garside, E.T. & J.S. Tait. 1958. Preferred temperature of rainbow trout (*Salmo gairdneri richardson*) and its unusual relationship to acclimation temperature. Can. J. Zool. 36: 563–567.

Gehlbach, F.R., C.L. Bryan & H.A. Reno. 1978. Thermal ecological features of *Cyprinodon elegans* and *Gambusia nobilis*, endangered Texas fishes. Tex. J. Sci. 30: 99–101.

Helwig, J.T. & K.A. Council. 1979. SAS user's guide. SAS Institute Inc., Raleigh, N.C. 83 pp.

Hill, L.G. & W.J. Matthews. 1980. Temperature selection by the darters *Etheostoma spectabile* and *Etheostoma radiosum* (Pisces: Percidae). Amer. Midl. Nat. 104: 412–415.

Hoar, W.S. 1955. Seasonal variations in the resistance of goldfish to temperature. Trans. Roy. Soc. Can. 49: 25–34.

Hutchison, V.H. & J.D. Maness. 1979. The role of behavior in temperature acclimation and tolerance in ectotherms. Amer. Zool. 19: 367–384.

Hynes, H.B.N. 1970. The ecology of running waters. University of Toronto Press, Toronto. 555 pp.

Kowalski, K.T., J.P. Schubauer, C.L. Scott & J.R. Spotila. 1978. Interspecific and seasonal differences in the temperature tolerance of stream fish. J. Therm. Biol. 3: 105–108.

Lynn, L.M. 1972. Seasonal changes in the benthic macroinvertebrates of Four Mile Creek (Butler County, Ohio). M.Sc. Thesis, Miami University, Oxford, Ohio. 57 pp.

Magnuson, J.J., L.B. Crower & P.A. Medvick. 1979. Temperature as an ecological resource. Amer. Zool. 19: 331–343.

Mathur, D., R.M. Schutsky & E.J. Purdy, Jr. 1982. Temperature preference and avoidance responses of the crayfish, *Orconectes obscurus*, and associated statistical problems. Can. J. Fish. Aquat. Sci. 39: 548–553.

Matthews, W.J. & J.T. Styron, Jr. 1981. Tolerance of headwater vs. mainstream fishes for abrupt physiochemical changes. Amer. Midl. Nat. 105: 149–158.

McCauley, R.W. & J.S. Tait. 1970. Preferred temperature of yearling lake trout, *Salvelinus namaycush*. J. Fish. Res. Board Can. 27: 1729–1733.

McCauley, R.W. & N.W. Huggins. 1979. Ontogenetic and non-thermal seasonal effects on thermal preferenda of fish. Amer. Zool. 19: 267–271.

Neel, J.K. 1951. Interrelations of certain physical and chemical features in a headwater limestone stream. Ecology 32: 368–391.

Precht, H., J. Christophersen, H. Hensel & W. Larcher. 1973. Temperature and life. Springer-Verlag, Berlin. 779 pp.

Reynolds, W.W. 1977. Temperature as a proximate factor in orientation behavior. J. Fish. Res. Board Can. 34: 734–739.

Reynolds, W.W. 1979. Perspective and introduction to the symposium: Thermoregulation in ectotherms. Amer. Zool. 19: 193–194.

Reynolds, W.W. & M.E. Casterlin. 1979. Behavioral thermoregulation and the 'final preferendum' paradigm. Amer. Zool. 19: 211–224.

Spotila, J.R., K.M. Terpin, R.R. Koons & R.L. Bonati. 1979. Temperature requirements of fishes from eastern Lake Erie and the upper Niagara River: a review of the literature. Env. Biol. Fish. 4: 281–307.

Stauffer, J.R. Jr., K.L. Dickson, J. Cairns, Jr. & D.S. Cherry. 1976. The potential and realized influences of temperature on the distribution of fishes in the New River, Glen Lyn, Virginia. Wildl. Monogr. 40: 1–40.

Sullivan, C.M. & K.C. Fisher. 1953. Seasonal fluctuations in the selected temperature of speckled trout, *Salvelinus fontinalis* (Mitchill). J. Fish. Res. Board Can. 10: 187–195.

Tyler, A.V. 1966. Some lethal temperature relations of two minnows of the genus *Chrosomus*. Can. J. Zool. 44: 349–364.

Winn, H.E. 1958. Comparative reproductive behavior and ecology of fourteen species of darters (Pisces:Percidae). Ecol. Monogr. 28: 155–191.

Originally published in Env. Biol. Fish. 11: 131–138

Morphological correlates of ecological specialization in darters

Lawrence M. Page & David L. Swofford
Illinois Natural History Survey, Champaign, IL 61820, U.S.A.

Keywords: Percidae, Habitat, Feeding, Reproduction, Canonical correlation

Synopsis

Darters feed on small benthic organisms, primarily insects, and evolutionarily have become increasingly small and benthic; most species are less than 80 mm in standard length. Constraints on decreasing body size include living in midwater and territoriality. Lineages of darters have arisen as new habitats were invaded. Consequently, members of different lineages often vary in characteristics correlated with specific habitat variables. While competition from established taxa undoubtedly has prevented additional habitat invasions, some darters appear to have overcome these barriers through feeding site diversification. Living in various habitats has lead to a variety of reproductive strategies, termed egg-burying, -attaching, -clumping, and -clustering. Sexually selective characteristics, correlated with type of reproductive behavior and habitat, often make the male more conspicuous and therefore are constrained by predation.

Introduction

Life cycles of organisms center about survival and reproduction; survival for darters primarily involves successfully occupying space (including avoiding predation and disease) and feeding. Ecological and morphological specializations of darters can be divided into those associated with feeding, habitat (space occupation), and reproduction. Specialization in one area sometimes constrains specialization in another, and some specializations may result in evolutionary artifacts having no apparent adaptive value.

The three genera of darters, *Percina, Ammocrypta,* and *Etheostoma,* contain about 150, or 20%, of the 750 freshwater fish species of North America north of Mexico. Darters often live in large populations and, in numbers of individuals, probably constitute more than 20% of the ichthyofauna. An ecological description that separates darters from most other groups of North American freshwater fishes, and therefore at least partly explains their success, is that they are small benthivores. The largest species reach about 150–170 mm in standard length (SL) – or about 175–200 mm in total length (TL) – but most are much smaller (Fig. 1). Because they are small and benthic they are able to capitalize very effectively on the enormous populations of benthos (primarily immature insects) in the freshwaters of North America. Minnows, the other highly speciose group (with about 220 species) of North American freshwater fishes, are mainly small midwater invertivores.

Primary lineages among darters (principally subgenera) have arisen as new habitats were invaded, and vary morphologically from one another primarily in characteristics correlated with specific habitat variables. Within most subgenera, habitat

David G. Lindquist & Lawrence M. Page (ed.), Environmental biology of darters. ISBN 90-6193-506-7

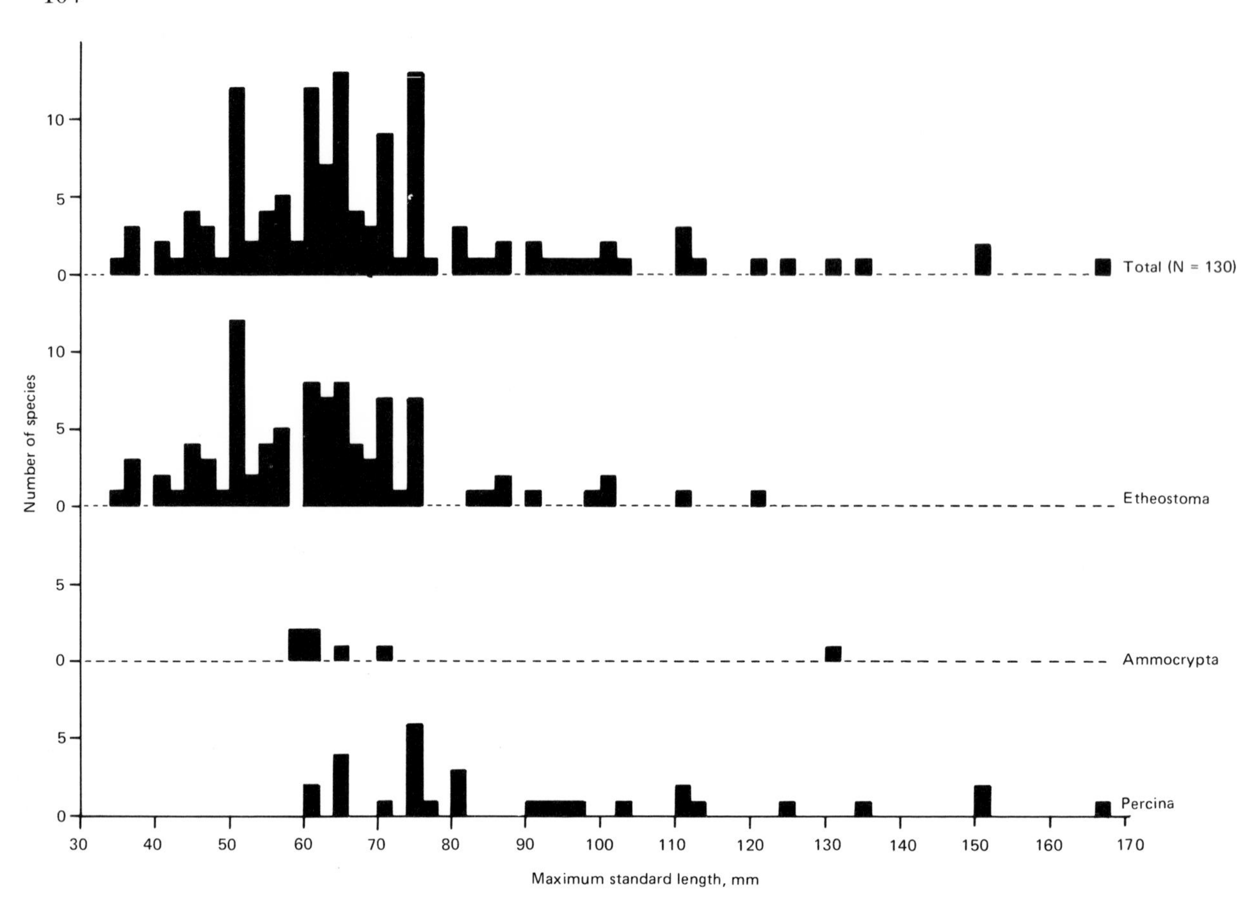

Fig. 1. Frequency distributions of maximum standard length, by genus and total.

preferences of the species are essentially the same – e.g. species of *Ammocrypta* (N = 6) live in sand runs, those of *Nothonotus* (N = 13) prefer rubble riffles, and those of *Boleichthys* (N = 10) live in swamps and sluggish streams. The most convincing evidence of the power of habitat-related selection on morphology is the fact that unrelated darters occupying the same habitat often resemble one another morphologically more than they resemble more closely related species.

The present diversity of darters is such that some unrelated groups overlap in habitat preferences. Resource partitioning among these groups (e.g. among coinhabiting riffle species) probably is accomplished through feeding diversification (Page 1983).

Morphological specializations other than those related to habitat and feeding are primarily characteristics which increase sexual dimorphism (i.e. are sexually selected). Darters have evolved a variety of reproductive strategies, and some of the most extreme morphological specializations are associated with reproductive behavior.

The following discussion addresses two questions:

1. How have darters specialized ecologically?;
2. What have been the morphological correlates of ecological specializations?.

While correlations between environmental and morphological variables do not necessarily imply cause-and-effect relationships, they at least reveal interesting patterns of covariance. Arguments implicating direct effects of environment on morphology (either through selection for adaptive traits or constraints on further evolution) are substantially reinforced when similar morphological characteristics are present in separate phyletic lines subjected to similar environmental pressures.

Because of the large number of variables which interact to define an organism's morphology and environment, attempts to find associations between them are likely to be hampered by the difficulty in interpreting the often complex relationships among the variables. While examinations of correlations between pairs of variables may be informative, hidden relationships may be missed because of correlations of both members of the pair with other variables, possibly masking more meaningful associations or leading to spurious correlations. By looking at correlations between *sets* of variables using multivariate techniques, the above problems can be alleviated (see, e.g. Harris 1975: 140). The canonical correlation between two sets of variables is the maximum product-moment correlation that can be obtained between linear combinations of the variables in each of the two sets. The resulting linear combinations (canonical vectors) can be interpreted so that the meaning of the relationship can be assessed in terms of the original variables. Following the discovery of the first pair of canonical vectors, additional pairs of functions can be found that maximize the correlation between new pairs of canonical variates, subject to the restriction that they be orthogonal to (uncorrelated with) all such previous pairs. This ability to discover independent relationships renders canonical correlation analysis extremely useful for examining the nature of morphological-habitat associations of darters. Unfortunately, diet and reproductive variables could not be analyzed similarly because too few data were avaliable.

Feeding specializations

Darters are first- and second-level carnivores which feed mainly on microcrustaceans (cladocerans, copepods, and ostracods) as juveniles and on immature aquatic insects (mostly midge and black flies, mayflies, and caddisflies) as adults (Page 1983). A second shift in diet, from small insects (mainly midge and black flies) to large insects (usually caddisflies and mayflies) and large crustaceans (amphipods, isopods, and crayfishes), is characteristic of large species (e.g. *P. caprodes* and *E. squamiceps* – Thomas 1970, Page 1974). Larger darters (species and individuals) eat a greater variety of, and larger, food items than do small darters. Very small species (e.g. *E. fonticola, E. proeliare,* and *E. microperca*) continue to ingest a high proportion of microcrustaceans as adults and seldom feed on insects larger than midge larvae (Schenck & Whiteside 1977, Burr & Page 1978, 1979).

Darters feed on small benthic organisms and they, in turn, have become increasingly small and benthic. Most (80%) reach a maximum SL (Fig. 1) of less than 80 mm (TL of less than 95 mm). A predator's body size should relate to prey size in a way that maximizes net energy gain per unit time spent foraging (Schoener 1971); in fishes, growth efficiency increases as predator size decreases relative to that of the prey (Kerr 1971). The large numbers of small benthic organisms (<5 mm long) available in a stream as food promote the evolutionary diminution of consumers, especially when the consumers are forced to forage among complex substrates (Miller 1979) such as rocky or debris-laden water bodies.

Reduction in size has been a pervasive aspect of darter evolution and has occurred among darters as a group, within genera, and within diverse subgenera. The largest darters are members of the

Table 1. Maximum standard lengths of darters categorized by genus, habitat, and water stratum.

Category	No. of species	Standard length, mm	
		x̄	SD
Genus			
Percina	30	91.8	29.1
Ammocrypta	7	71.7	26.0
Etheostoma	93	62.2	16.3
Total	130	69.5	23.4
Habitat			
Gravel run	26	93.9	23.6
Riffle	53	70.6	22.4
Sand run	7	61.0	4.8
Pool	44	56.9	13.0
Stratum			
Midwater	29	95.2	28.2
Benthic	101	62.8	15.6

genus *Percina* (x̄ maximum SL = 91.8 mm, Table 1), and the smallest are members of the advanced genus *Etheostoma* (x̄ = 62.2 mm); species of *Ammocrypta* are intermediate phylogenetically and in size (x̄ = 71.7 mm). Intrasubgeneric reductions (Fig. 2) tend to obsure intrageneric reduction, but the trend is evident in the largest genus, *Etheostoma* (Fig. 2). The smallest species of *Etheostoma* (~35 mm SL) are among the smallest North American freshwater fishes (Lee et al. 1980).

Although the tendency at all taxonomic levels has been for darters to become smaller, reduction in body size is constrained by habitat and water stratum preferences (Table 1). Gravel-run species are the largest, pool species are the smallest (Fig. 3); riffle and sand-run species are intermediate in size (an interesting exception is *E. tippecanoe,* a riffle species reaching only 35 mm SL). Midwater species are, in general, larger than benthic species (Fig. 3).

Reduction among darters has been carried to its extreme in benthic pool species. Smaller size increases vulnerability to a greater number of predators, but diminution in benthic darters also facilitates their concealment among components of the substrate. Midwater darters, more visible and exposed to predators, have remained relatively large and vulnerable to fewer predators. Sand-run darters (species of *Ammocrypta* and *E. vitreum*) are benthic, intermediate in size and occupy a simple substrate, but reduce exposure to predators through cryptic coloration (transparency) and by burrowing into the substrate. It may be that insufficient time has elapsed for darters to become even smaller than they have; the lower limit to body size in teleosts may be determined by gametogenesis, a process which seems to require a body size of at least 10 mm (Bruun 1940).

Sexual dimorphism in size probably is universal in darters and, in all but two subgenera, males are larger than females (Page 1981). In territorial animals it is usual for large males to be more successful than small males in holding territories and ultimately in reproducing (e.g. Estes 1974); the only darters thought to be non-territorial are species of the subgenera *Ozarka* and *Boleichthys* of *Etheostoma* (Strawn 1956, Distler 1972, Burr & Page 1978), the two groups in which females are larger than males. The selective pressure on males to be large and defend a territory conflicts with the tendency among darters to become smaller and more efficient consumers.

Each genus of darters has a small proportion of species much larger than the mode (Fig. 1). Size-

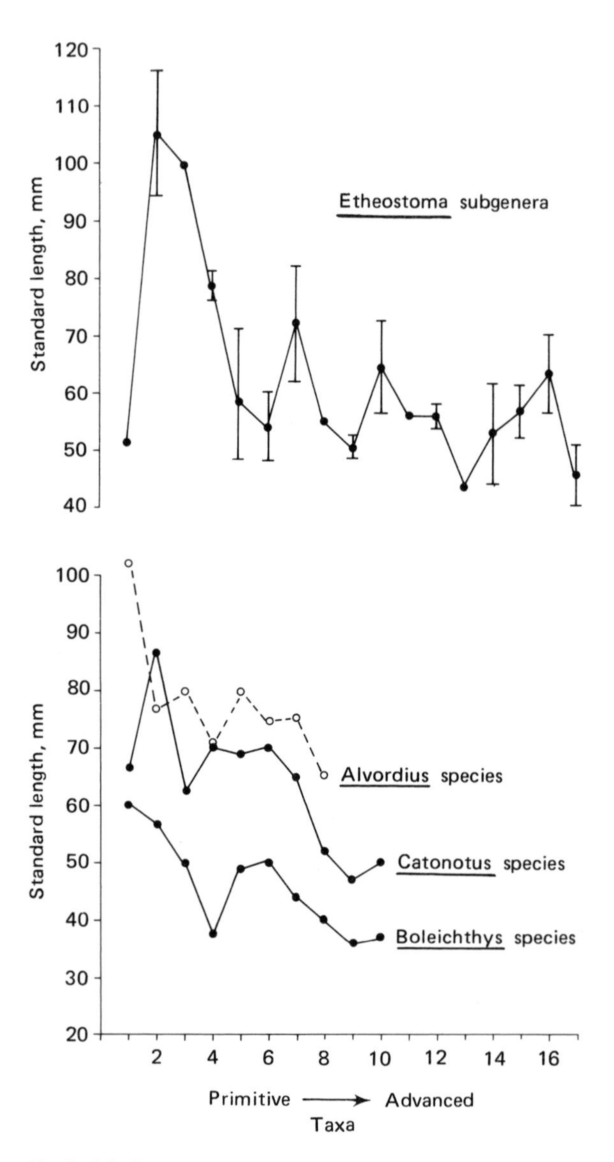

Fig. 2. Maximum standard lengths of subgenera of *Etheostoma* (subgeneric mean ± 2 SE) and maximum standard lengths of species in the subgenera *Alvordius* (of *Percina*) and *Catonotus* and *Boleichthys* (of *Etheostoma*). Taxa are arranged in an approximate sequence from primitive to advanced (Page 1981).

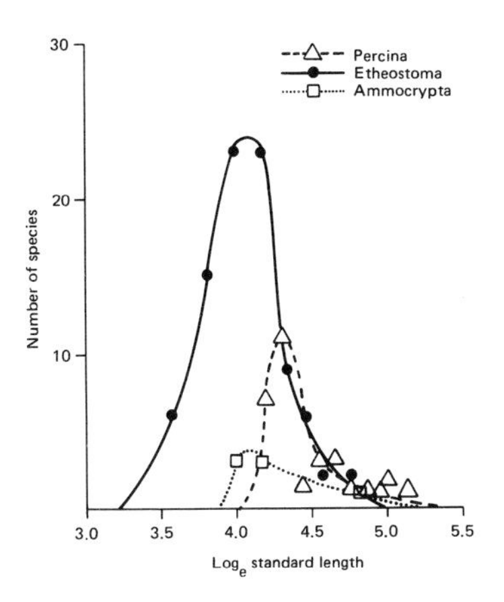

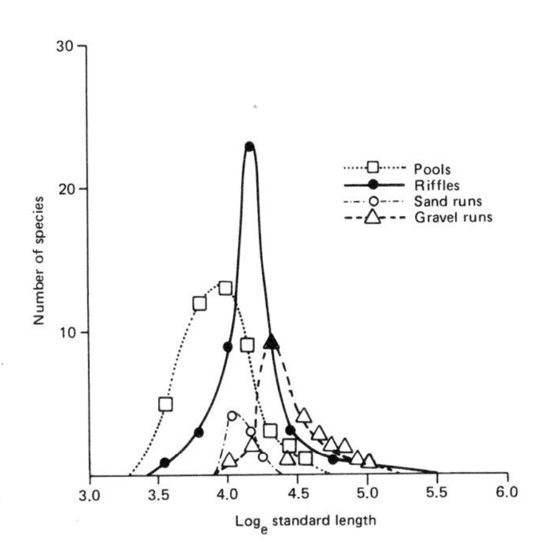

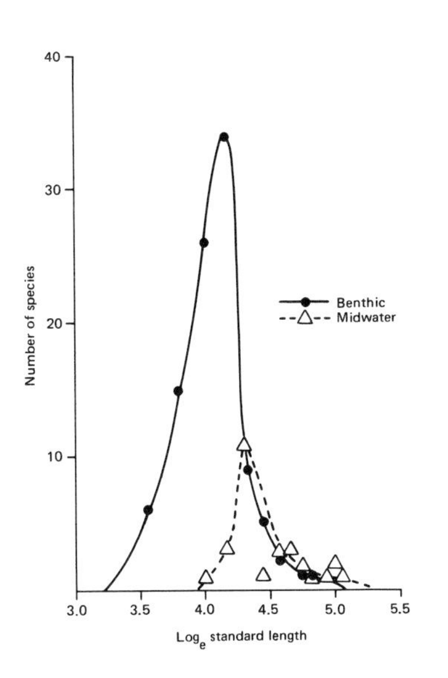

Fig. 3. Length frequency distributions by genus (top), habitat (middle), and water stratum (bottom). Symbols represent size-class means.

frequency distributions of many taxa show similar right-skewness (Hemmingsen 1934), and may indicate a common evolutionary regulating mechanism as yet unidentified (Kirchner et al. 1980).

Some characteristics of darters may be artifacts of reduced body size and thus may be adaptively neutral (Gould & Lewontin 1979, Jaksic 1981). Interruptions in the head canals of darters (Page 1977) occur only in small species (Fig. 4) and may result from incomplete closures (truncated development) of the canals. The infraorbital canal grows posteriorly from just behind the nostril and anteriorly from its junction with the lateral canal behind the eye (Collette 1962); in small darters, development stops before the two halves fuse under the eye to form an uninterrupted canal. Similarly, failure of the right and left halves of the supratemporal canal to meet medially in small darters leaves an interruption. However, most small

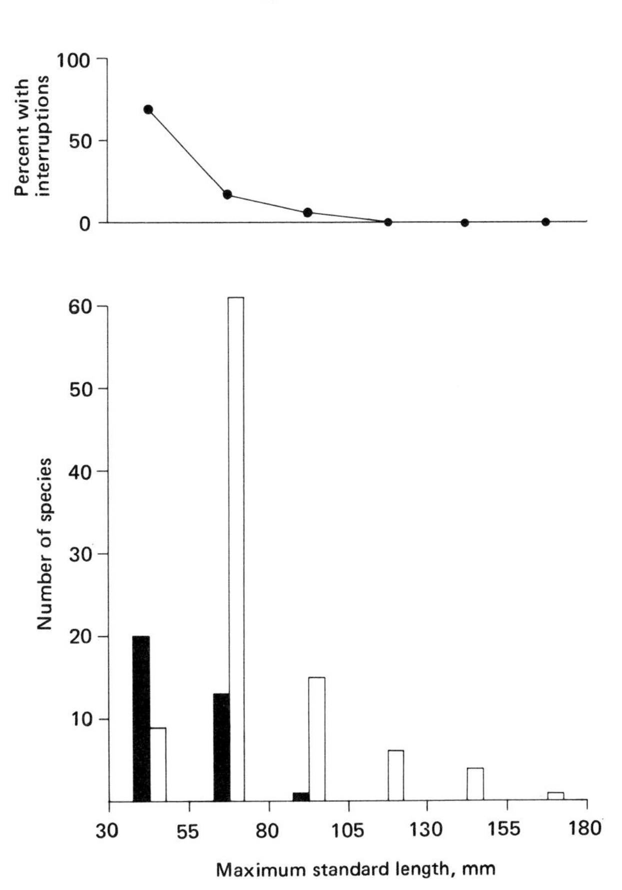

Fig. 4. Length frequency distribution of darters with (dark bars) and without (open bars) head canal interruptions.

darters occupy slow or standing water and it may be that interruptions in the lateralis system actually enhance sensory perception in these species by exposing neuromasts that otherwise are within canals. (This point is developed further in the section on habitat specializations.)

As in diminution, the tendency among darters to become benthic has been pervasive; darters live among the organisms on which they feed, thereby increasing the efficiency of foraging. Probably no darter occupies a water stratum far off the bottom, in contrast to minnows which stratify at all levels of the water column (Mendelson 1975, Baker & Ross 1981, Surat et al. 1982). The distinction made between 'midwater' and 'benthic' darters (Fig. 3) is between those which predominantly swim near (but off the bottom) and those which predominantly sit on the bottom.

Most species of *Percina* are midwater forms, although some definitely are benthic (e.g. *P. roanoka* and *P. copelandi*), and others are sufficiently intermediate to make categorization somewhat difficult (e.g. *P. evides*). Almost all species of *Ammocrypta* and *Etheostoma* are benthic; *E. nianguae* and *E. sagita* may be intermediate (Pflieger 1978, pers. obs.).

Advanced species of darters have become part of the benthos and, in the process, have lost the swimbladder. In 17 species of darters examined (Page 1983), a large swimbladder was found only in the highly midwater *P. cymatotaenia* (swimbladders in three specimens averaged 26% SL), and moderate swimbladders were found in *P. sciera*, *P. phoxocephala*, and *P. maculata* ($\bar{x}$ = 15–16% SL). A small swimbladder ($\bar{x}$ = 8% SL) was found in one of three adult *P. evides* examined; the other two *P. evides* and all other species examined (*P. palmaris, P. roanoka, P. copelandi, P. shumardi, P. ouachitae, P. antesella, A. pellucida, E. nianguae, E. zonale, E. nigrum, E. vitreum,* and *E. caeruleum*) lacked swimbladders.

Miller (1979) categorizes small teleosts as nektonic, epibenthic, or cryptobenthic on the basis of their adult behavior and trophic niche. These categories fit darters well; their transition from midwater (nektonic) to benthic has been both to epibenthic and to cryptobenthic life styles. Most darters (including most riffle species) have become epibenthic but others, when permitted by habitat characteristics, have become cryptobenthic. In the latter group of species, evolution has exploited the availability of physical barriers and behaviorly placed them between the darters and potential predators; e.g. *E. flabellare* and *E. squamiceps* hide under large stones (Page 1974) and have accomodating dark coloration. Species of *Ammocrypta* and *E. vitreum* live in sand runs, have cryptic coloration (transparency), and bury in the sand with only their eyes and snout protruding (Forbes & Richardson 1908, Winn & Picciolo 1960, Lee & Ashton 1979, Trautman 1981); correspondingly, their eyes are located more dorsally than those of other darters.

Other benthic-associated characteristics of darters (e.g. a flattened venter) are discussed below in relation to habitat preference.

Rafinesque was impressed by the striking variation in mouths of the darters known to him (*P. caprodes, E. blennioides, E. flabellare*) and coined for them the name, *Etheostoma,* meaning 'various mouths' (Jordan & Evermann 1896). The tremendous variation in mouth size and position, in contrast to the great redundancy in food items (especially fly larvae) found in darters (Page 1983), suggests that dietary separations among syntopically occurring species are based on where darters feed, rather then what they feed on. In his study on five species of *Etheostoma,* Wehnes (1973) concluded that subterminal mouths are adaptive for foraging on the tops of rocks, and terminal mouths are adaptive for foraging on the sides of rocks. Few other field observations on feeding have been recorded, but extrapolations from mouth morphologies suggest some feeding behaviors. Pointed-snout species with terminal mouths, e.g. *P. phoxocephala* and *E. caeruleum,* extract food organisms from among small stones. Species with blunt snouts and subterminal months, e.g. species of the subgenus *Nanostoma,* feed on the sides and tops of boulders. Upturned mouths, as in *E. flabellare* and *E. squamiceps,* facilitate removal of organisms from the undersides of large stones. The long bulbous snout of *P. caprodes* and other logperches is used to expose food organisms by flipping stones and debris (Keast & Webb 1966) and by rooting through gravel (pers. obs.). Future studies on feeding habits of darters will be more informative if they determine precisely where darters are feeding as well as the composition of the diet.

Habitat specializations

Canonical correlation

Canonical correlation analysis followed Cooley & Lohnes (1971) and was implemented using the BMDP6M program from the BMDP package (Dixon et al. 1981). After derivation of the canonical vectors, correlations of the original variables with these vectors ('loadings') were computed. The resulting 'canonical structure' served as the basis for biological interpretation.

Bartlett's test for the equality of eigenvalues was used to test the significance of canonical correlations; a pair of canonical vectors was considered important only if a test of equality of the remaining eigenvalues (squared canonical correlation coefficients) was significant at $P < 0.01$. Separate analyses were performed for each of the three sets of morphological variables (meristic, miscellaneous, and mensural), correlating each set with the full set of habitat variables.

The 78 species examined included the 58 species in Table 5 (excluding *E. aquali,* the red snubnose darter, and the lowland snubnose darter) and 20 additional species representing a diversity of morphology (*P. roanoka, P. cymatotaenia, P. sciera, P. shumardi, P. uranidea, P. phoxocephala, P. carbonaria, P. nasuta, P. nigrofasciata, A. beani, A. asprella, A. clara, E. sellare, E. blennius, E. cinerium, E. sagitta, E. mariae, E. tuscumbia, E. luteovinctum,* and *E. australe*); all subgenera (Page 1981) were represented.

Five habitat and 66 morphological variables were scored for each species. We chose to treat species rather than individuals as observations in the analysis because many of the morphological variables are constant within species, and for many species little information is available on intraspecific (including age-specific) variation in habitat preference. Because of these considerations, variables were scored for adult darters.

Assignments of values to variables were based on published information, reviewed by Page (1983), and on personal observations. (We have had field experience with all of the 78 species considered except *E. sellare.*) Meristic and mensural data were from Page (1981). A list of the variables used in the analysis follows; a complete listing of the data matrix will be provided on request.[1]

Habitat variables: 1. Substrate (SUB): 1 = mud, 2 = mud-sand, 3 = sand, 4 = sand-gravel, 5 = gravel, 6 = gravel-rubble, 7 = rubble, 8 = rubble-bedrock, 9 = bedrock. 2. Current (CUR): 1 = none, 2 = none-slow, 3 = slow, 4 = slow-fast, 5 = fast. 3. Depth (DEP): 1 = shallow, less than 25 cm; 2 = deep, more than 25 cm. 4. Size of stream (STR): 1 = headwater, less than 1 m wide; 2 = creek, 1–5 m; 3 = small river, 5–25 m; 4 = medium river, 25–50 m; 5 = large river, more than 50 m; 6 = standing water. 5. Water stratum occupied (STM): 1 = benthic, 2 = benthic-midwater, 3 = midwater.

Morphological variables: 1. Mean number of modified scales (enlarged and strongly toothed) in midbelly row (NMS). 2. Mean number of lateral scales (NLS). 3. Mean percent of scales in lateral line row that are pored (PER). 4. Mean number of scale rows above lateral line (NSA). 5. Mean number of scale rows below lateral line (NSB). 6. Mean number of scales around caudal peduncle (NSCP). 7. Mean number of pored scales on caudal fin (NPSC). 8. Mean number of spines in first dorsal fin (NDS). 9. Mean number of rays in second dorsal fin (NDR). 10. Mean number of spines in anal fin (NAS). 11. Mean number of rays in anal fin (NAR). 12. Mean number of rays in pectoral fin (NPR). 13. Mean number of transverse scale rows (NTS). 14. Modal number of infraorbital pores (NIOP). 15. Modal number of preoperculomandibular pores (NPOM). 16. Modal number of branchiostegal rays (NBR). 17. Modal number of supratemporal pores (NSTP). 18. Breast squama-

[1] Readers who question the mixture of binary, ordinal, and continuous variables in our multivariate analysis are referred to the discussion on this subject by Harris (1975).

tion of males (BRSM): 1= fully scaled, 2= reduced squamation, 3= unscaled. 19. Breast squamation of females (BRSF): 1= fully scaled, 2 = reduced squamation, 3 = unscaled. 20. Anterior belly squamation of males (ABS): 1= unscaled, 2 = scaled. 21. Belly midline squamation of males (BMSM): 1 = complete or incomplete rows of unmodified scales, 2 = scaleless strip(s) from pelvic fins to genital papilla, 3 = with a midbelly row of modified scales on which the longest tooth is always less than 40% of the entire length of the modified scale, 4 = with a midbelly row of modified scales on which the teeth are sometimes longer than 40% of the entire length of the scales. 22. Belly midline squamation of females (BMSF): 1 = fully scaled, 2 = variable, from almost fully scaled to a naked strip extending one-half distance from genital papilla to pelvic fins, 3 = variable, from almost unscaled to a naked strip extending one-half distance from genital papilla to pelvic fins, 4 = unscaled from pelvic fins to genital papilla. 23. Flesh when alive (FLES): 1 = opaque, 2 = translucent. 24. Pigment bar on cheek as in *E. virgatum* (see photo in Kuehne & Small 1971) (PIG): 1 = absent, 2 = present. 25. One medial black basicaudal spot (SPOT): 1 = absent, 2 = present. 26. Black midlateral stripe, including strongly interconnected blotches (STRI): 1= absent, 2 = present. 27. Vertical bars on side (BAR): 1= absent, 2 = present. 28. Red/orange band(s) in dorsal fin(s) (BAND): 1= absent, 2 = present. 29. Blue/green on body (BLG): 1 = absent, 2 = present. 30. Red/orange on body (RED): 1= absent, 2 = present. 31. W/X marks on side (W/X): 1= absent, 2 = present. 32. Large dorsolateral saddles (SAD): 1 = absent, 2 = present. 33. Conical snout projecting well beyond premaxilla (CON): 1= absent, 2 = present. 34. Premaxillary frenum (FREN): 1 = absent or extremely narrow, 2 = present and not extremely narrow. 35. Branchiostegal membrane connection (BMC): 1 = membranes not or slightly joined, 2 = membranes narrowly to moderately joined, 3 = membranes broadly joined. 36. Lateral line (ARCH): 1 = straight or slightly arched, 2 = arched distinctly upward anteriorly. 37. Breeding tuberculation (including tubercular ridges) (TUB): 1 = absent, 2 = present on males only, 3 = present on males and females. 38. Infraorbital canal (ICAN): 1 = uninterrupted, 2 = interrupted. 39. Supratemporal canal (SCAN): 1 = uninterrupted, 2 = interrupted. 40. Palatine teeth (PAL): 1 = present, 2 = absent. 41. Prevomerine teeth (PVO): 1 = present, 2 = absent. 42. Pelvic fin flaps (FLAP): 1 = absent, 2 = present. 43. Dorsal fin knobs (KNOB): 1 = absent, 2 = present. 44. Wide flattened genital papilla on female (WIDE): 1 = absent, 2 = present. 45. Swollen flesh on head and nape of breeding male (SWOL): 1 = absent, 2 = present. 46. Adult females average smaller (= 1) or larger (= 2) than adult males (DIM). 47. Maximum standard length recorded for species, mm (MAX). 48. Mean standard length, mm (SL). 49. Mean head length, mm (HL). 50. Mean head width, mm (HW). 51. Mean body depth, mm (BD). 52. Mean

Table 2. Correlations of habitat and meristic variables with first three pairs of canonical variates.[a]

Variable	Canonical vector		
	I	II	III
Habitat			
SUB	0.200	**0.920**	0.310
CUR	**0.312**	**0.318**	**0.873**
STR	**0.947**	−0.110	−0.231
DEP	**0.339**	**−0.687**	**−0.416**
STM	**0.421**	0.032	0.098
Meristic			
NMS	**0.843**	−0.136	−0.046
NLS	**0.716**	−0.051	0.067
PER	**0.544**	0.152	**0.375**
NSA	**0.705**	**0.314**	0.058
NSB	**0.772**	0.201	−0.136
NSCP	**0.579**	−0.068	0.264
NPSC	**0.482**	−0.138	0.177
NDS	**0.768**	0.114	0.272
NDR	**0.400**	**0.414**	−0.105
NAS	**0.256**	**0.395**	0.185
NAR	**0.604**	−0.077	−0.071
NPR	**0.429**	0.216	**0.343**
NTS	**0.737**	0.253	−0.061
NIOP	**0.393**	−0.019	**0.453**
NPOM	**0.312**	0.202	**0.346**
NBR	0.150	**−0.443**	0.139
NSTP	**−0.453**	−0.065	−0.147
Canonical R	0.929	0.736	0.726

[a] Loadings emphasized in the text are shown in boldface type.

predorsal length, mm (PDL). 53. Mean caudal peduncle depth, mm (CPD). 54. Mean first dorsal fin base length, mm (D1BL). 55. Mean second dorsal fin base length, mm (D2BL). 56. Mean pectoral fin length, mm (P1L). 57. Mean snout length, mm (SNL). 58. Mean interpelvic fin width, mm (IPW). 59. Mean pelvic fin base length, mm (P2BL). 60. Mean interorbital width, mm (IOW). 61. Mean gape width, mm (GW). 62. Mean anal fin length, mm (AL). 63. Mean second dorsal fin length, mm (D2L). 64. Mean distance from origin of second dorsal fin to center of caudal base, mm (QL). 65. Mean pelvic fin length, mm (P2L). 66. Mean caudal fin length, mm (CFL).

Habitat and meristic variables

The canonical structure for the first three pairs of canonical variates is shown in Table 2. The first pair of canonical variates identifies a tendency for midwater darters in medium to large streams to have high meristic counts (but only 3 supratemporal pores). The strength of the canonical correlation is evident from the plot shown in Figure 5, in which the species have been located in the space determined by the first pair of canonical vectors. Species with high scores on both linear combinations include *P. carbonaria, P. caprodes, P. phoxocephala, P. nasuta, P. maculata, P. sciera,* and *P. cymatotaenia,* and represent five different subgenera (*Percina, Swainia, Alvordius, Hadropterus, Odontopholis*).

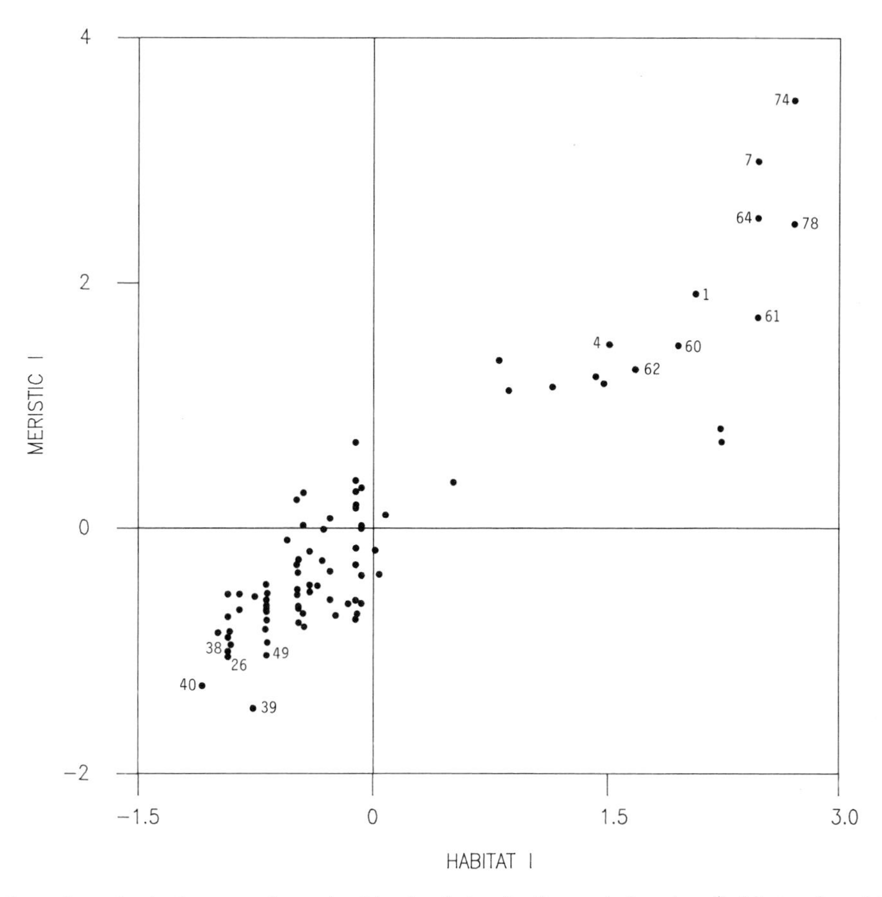

Fig. 5. Locations of species in the space determined by the first pair of canonical vectors (habitat and meristic variables). Species identified are *Percina carbonaria* (74), *P. caprodes* (7), *P. phoxocephala* (64), *P. nasuta* (78), *P. maculata* (1), *P. sciera* (61), *P. cymatotaenia* (60), *P. evides* (4), *P. shumardi* (62), *Etheostoma kennicotti* (49), *E. proeliare* (38), *E. chlorosomum* (26), *E. microperca* (40), and *E. fonticola* (39).

Meristic counts seem to be a function of body size: small darters have fewer scales and fin rays. The alternative of having similar numbers of proportionally smaller scales and rays seems rarely to have occurred in darters, as evidenced by high correlations of meristic counts with maximum standard length (Fig. 6). The strongest correlations ($r>0.5$, $P<0.0001$) were found with mean numbers of lateral scales ($r = 0.761$), transverse scale rows ($r = 0.737$), scales below ($r = 0.714$) and above ($r = 0.702$) the lateral line, pored scales on caudal fin ($r = 0.625$), dorsal spines ($r = 0.597$), scales around the caudal peduncle ($r = 0.588$), anal rays ($r = 0.581$), pectoral rays ($r = 0.543$), and dorsal rays ($r = 0.530$). Combining dorsal spines and rays into a single variable (total dorsal fin elements) in-

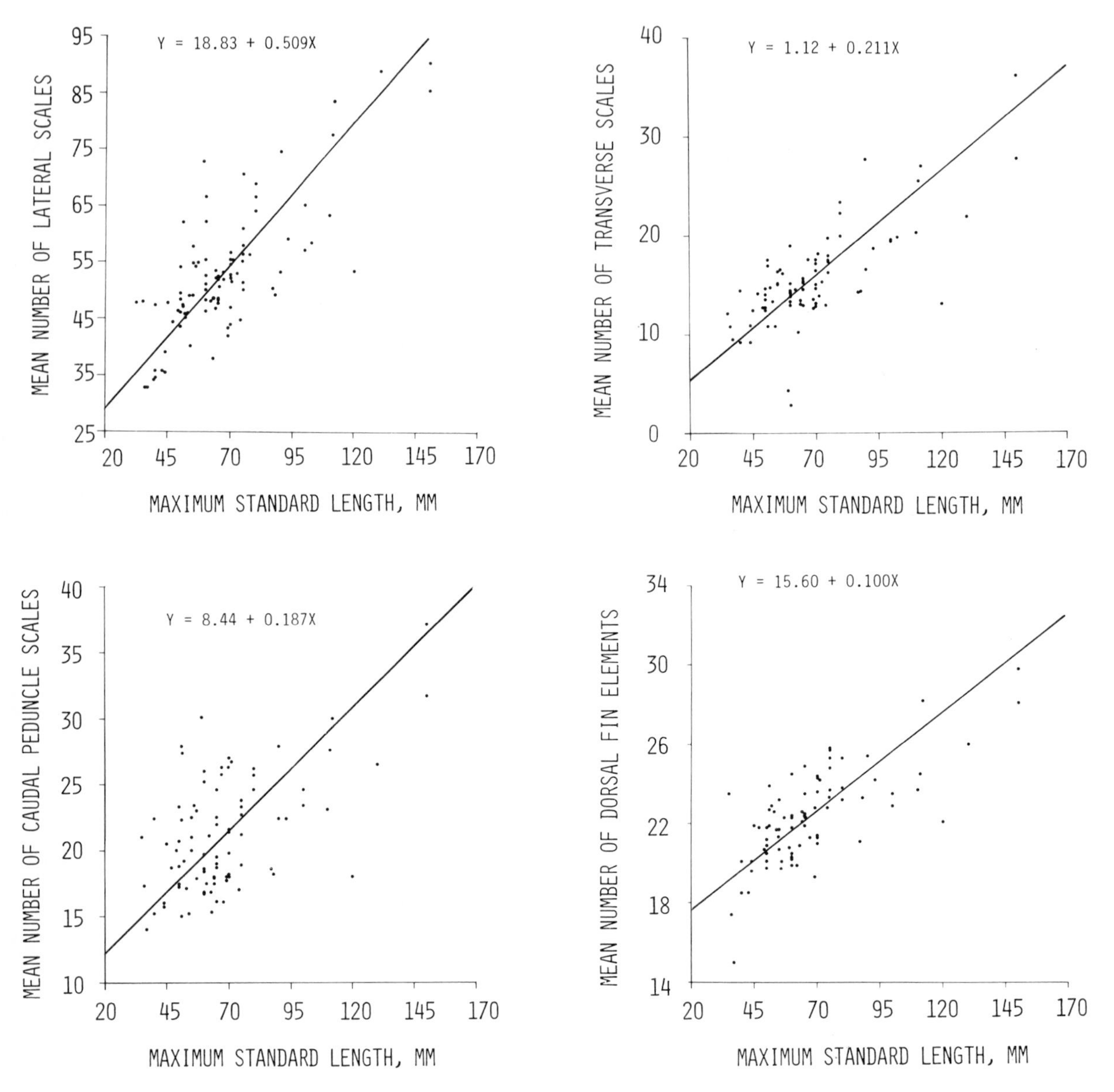

Fig. 6. Plots illustrating effect of length on meristic counts. Maximum standard lengths (given by Page 1983) were used rather than an estimate of mean adult size due to the ambiguity of the latter in organisms having indeterminate growth. Because both variables in each plot are subject to natural variability, linear regressions were computed using the reduced major axis method, a model II technique (= GM-regression; Ricker 1973).

creased the correlation coefficient to 0.707, indicating that variation in one is dependent on the other. Lowest correlations were found for anal spines ($r = 0.123$, $P = 0.28$) and branchiostegal rays ($r = 0.163$, $P = 0.16$).

We do not believe that some darters have increased in size as a result of their occurrence in larger streams. Rather, we believe that the earliest darters were large and those species occupying midwater strata of larger streams have remained large due to constraints placed upon body size by characteristics of their habitat, as discussed above. Species with the lowest scores on the first pair of canonical vectors are among the most highly evolved darters (Bailey & Gosline 1955, Page 1981): *E. fonticola, E. microperca,* and *E. proeliare* (subgenus *Boleichthys*); *E. chlorosomum* (*Vaillantia*); and *E. kennicotti (Catonotus)*.

The second pair of canonical variates suggests that darters in shallow, rocky habitats have elevated numbers of dorsal rays and anal spines but fewer branchiostegal rays. Scoring highest on both pairs of canonical vectors were members of the subgenus *Nanostoma,* which along with *E.* (*Boleichthys*) *microperca* are the only darters with five branchiostegal rays (all others have six). The lowest scores were for species of *Ammocrypta,* which are among the few darters having one, rather than two, anal spines. No adaptive significance of these characteristics is apparent, and the second canonical correlation appears to have little ecological significance.

The third pair of canonical variates highlights a tendency for darters in slow, deep water to have a lower percentage of pored scales in the lateral line, fewer infraorbital and preoperculomandibular pores, and fewer pectoral rays. Generally, the lateralis system is well-developed in fishes which swim continuously or are otherwise exposed to currents but is reduced in fishes which live in sluggish water or otherwise avoid currents (Lowenstein 1957). A well-developed system of canals to protect the neuromasts from continuous stimulation, and ultimately sensory fatigue (Lowenstein 1957), is necessary in fast-moving water but apparently not in slow-flowing water. Interruptions in canals may actually enhance sensitivity by exposing neuromasts otherwise contained within canals.

In contrast to the trend for sluggish water species, darters in fast, shallow water tend to have a higher number of pectoral rays; this presumably relates to an increase in the size of the fin (see the 'Habitat and mensural variables' section).

Habitat and miscellaneous variables

Three significant relationships were detected involving linear combinations of habitat and miscellaneous variables. Correlations of the original variables with these canonical vectors are shown in Table 3. The first pair of canonical variates identifies a trend in which midwater darters occurring over rocky substrates are large, tend to have scales on the belly of the male, a basicaudal spot, a midlateral stripe, a conical snout, and males exceeding females in size. In contrast, benthic darters living over sand or mud are small, have no basicaudal spot or midlateral stripe, those living on sand have no scales on the belly, and those living over mud (i.e. species of the subgenus *Boleichthys* and *E. cragini*) tend to have females larger than males. We do not understand the evolutionary significance of losing scales on the portion of the body proximal to sand (but not gravel or mud), but the phenomenon is carried to an extreme in species of the subgenus *Ammocrypta,* which bury in sand. The most highly evolved species of *Ammocrypta* have lost scales over most of the body (Williams 1975). Species of the subgenus *Imostoma* are the most sand-inhabiting species of *Percina* and are also the only species of the genus to have secondarily undergone reduction in the midbelly row of modified scales. (These scales are often absent, as are the other belly scales on species of *Imostoma*).

A midlateral stripe and basicaudal spot enhance crypticity of midwater species by disrupting the body outline and serving as a false eyespot, respectively. Unlike midwater darters which are likely to be viewed laterally by potential predators, benthic darters are more often viewed dorsally, and lack these characteristics. The only species of darters in which females are larger than males are those few thought to be nonterritorial (i.e. species of the subgenera *Boleichthys* and *Ozarka,* as discussed

Table 3. Correlations of habitat and miscellaneous variables with first three pairs of canonical variates.

Variable	Canonical vector		
	I	II	III
Habitat			
SUB	**0.319**	**0.685**	0.572
CUR	0.114	**0.893**	− 0.209
STR	**0.935**	− 0.129	− 0.324
DEP	0.182	− **0.732**	− 0.541
STM	0.222	0.006	− 0.172
Miscellaneous			
BRSM	0.101	0.129	0.075
BRSF	0.113	0.184	0.050
ABS	− 0.162	0.129	0.286
BMSM	**0.707**	− 0.077	− 0.441
BMSF	0.213	− 0.196	− 0.408
FLES	− 0.243	− 0.043	− 0.378
PIG	0.002	− 0.066	0.486
FREN	0.149	**0.309**	− 0.057
BMC	− 0.171	0.248	0.195
ARCH	− 0.286	− **0.537**	− 0.095
TUB	− 0.144	− 0.072	− 0.377
ICAN	− 0.221	− 0.199	0.400
SCAN	− 0.340	− 0.195	0.193
PAL	− 0.141	0.138	0.210
PVO	− 0.109	0.021	0.147
CON	**0.428**	0.010	− 0.178
FLAP	− 0.184	− **0.362**	− 0.054
KNOB	− 0.080	− 0.186	0.377
WIDE	− 0.092	− 0.015	0.589
SWOL	− 0.066	− 0.149	0.628
DIM	− 0.331	− **0.455**	− 0.174
SPOT	**0.477**	− **0.528**	− 0.153
STRI	**0.421**	− 0.122	0.133
BAR	0.130	**0.300**	− 0.118
BAND	− 0.098	0.152	0.224
BLG	− 0.094	**0.379**	0.212
RED	− 0.205	**0.452**	0.100
X/W	− 0.180	− **0.381**	0.149
SAD	− 0.087	0.291	− 0.200
MAX	**0.641**	0.167	− 0.030
Canonical R	0.935	0.927	0.906

above). The fact that most of these species live on mud may be related to their lack of territoriality, and consequently to their altered female:male size ratio.

The second pair of canonical variates suggests that darters in fast, shallow, rocky streams have bright colors and vertical bars but no X/W marks on the side, a straight lateral line, males larger than females, a premaxillary frenum, and lack pelvic fin flaps and a basicaudal spot.

Bright colors may help conceal darters living in riffles; many species living among the colored stones have hues of red, yellow, and blue, and those living in algae (e.g. *E. blennioides, E. zonale*) are green. It may also be that living in shallow riffles reduces predation to a level at which selection for crypticity is no longer a dominating influence on pigmentation, and the epigamic advantages of bright colors prevail. Males of most riffle darters are colored more brightly than females. Riffle darters, benthic and likely to be seen dorsally or dorsolaterally by potential predators, have vertical bars to disrupt the outline of the body. Bars are common on deep-bodied benthic fishes; horizontal stripes are common on elongate midwater fishes (Barlow 1972).

Many quiet-pool species are drably colored and speckled with black or brown pigment, often in the form of X- or W-marks, and thus blend in well with the sand or debris-laden substrate of quiet pools.

Darters in slow, deep water tend to have an arched lateral line (i.e. species of *Boleichthys*). Other sluggish water fishes also have a curved rather than straight lateral line, although it may be convex (e.g. species of *Lepomis*), or concave (e.g. *Notemigonus crysoleucas*) relative to the dorsum. A curved lateral line has an increased vertical dimension on the side of the fish, and thus may have an enhanced receptivity that works in sluggish, but not fast, water.

The presence or absence of a premaxillary frenum is presumably associated with the mode of feeding, and consequently the type of habitat. Species living in riffles and gravel runs have a well-developed frenum; those living in pools (e.g. *E. nigrum*) and sand runs (*Ammocrypta* spp.) often have at most a strongly reduced frenum, and the premaxillary is freely protrusile. The protrusility of the premaxillary may provide additional maneuverability to the pool and sand-run species which have to remove their food more carefully from less stable substrates.

A plot showing the locations of each of the 78 species in the space defined by the second pair of

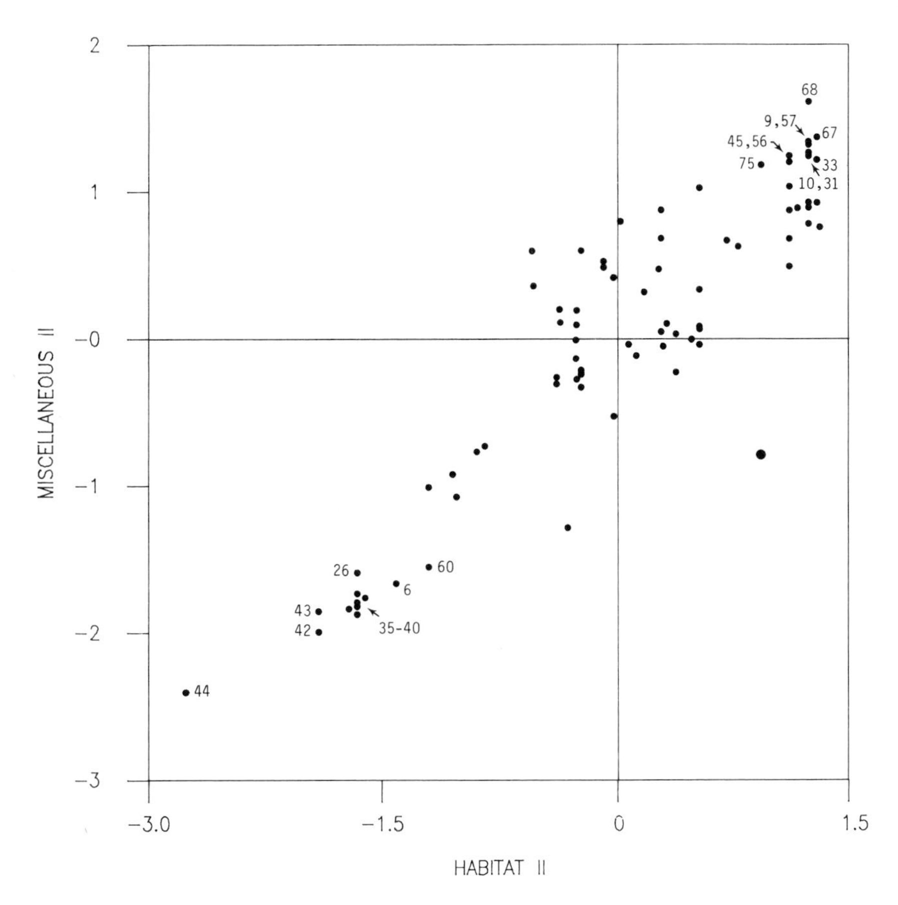

Fig. 7. Locations of species in the space determined by the second pair of canonical vectors (habitat and miscellaneous variables). Species identified are *Etheostoma blennius* (68), *E. sellare* (67), *E. tetrazonum* (9), *E. microlepidum* (57), *E. variatum* (10), *E. lepidum* (31), *E. grahami* (33), *E. longimanum* (45), *E. maculatum* (56), *Percina nigrofasciata* (75), *P. cymatotaenia* (60), *P. copelandi* (6), *E. chlorosomum* (26), *E. exile* (35), *E. fusiforme* (36), *E. gracile* (37), *E. proeliare* (38), *E. fonticola* (39), *E. microperca* (40), *E. olmstedi* (43), *E. nigrum* (42), and *E. perlongum* (44).

canonical vectors is shown in Figure 7. This plot is especially interesting because of the diversity of subgenera represented in the upper right quadrant (high scores on both canonical vectors). In addition, most members of the subgenus *Boleosoma* have low scores on this pair of canonical variates (lower left quadrant), but *E. longimanum,* the only species of *Boleosoma* included which inhabits rocky riffles, is found with other riffle darters at the opposite extreme.

Little ecological significance could be attributed to the third canonical correlation, despite its relatively high magnitude. Cooley & Lohnes (1971) and Pimentel (1979) have observed that the number of canonical correlations that are statistically significant may exceed the number that are biologically meaningful.

Habitat and mensural variables

The canonical structure for the final comparison is presented in Table 4. The first pair of canonical variates reiterates a pattern evident from the comparison of meristic and habitat variables: midwater darters in large streams tend to be large (i.e. all mensural variables have large positive loadings as do current, stratum, and stream-size variables).

The second pair of canonical variates suggests that benthic darters in fast, shallow rocky habitats have deep caudal peduncles, deep bodies, long

Table 4. Correlations of habitat and mensural variables with first three pairs of canonical variates.

Variable	Canonical vector		
	I	II	III
Environmental			
SUB	**0.454**	**−0.666**	**−0.479**
CUR	**0.669**	**−0.619**	0.153
STR	**0.935**	**−0.607**	−0.119
DEP	0.182	**−0.706**	**0.608**
STM	**0.404**	−0.286	**0.511**
Mensural			
SL	**0.654**	0.145	−0.118
HL	**0.688**	0.098	−0.238
HW	**0.651**	−0.121	−0.193
BD	**0.694**	**−0.198**	**−0.350**
PDL	**0.740**	0.101	−0.109
CPD	**0.628**	**−0.245**	**−0.418**
D1BL	**0.802**	0.006	−0.147
D2BL	**0.484**	−0.118	−0.201
P1L	**0.530**	**−0.264**	−0.141
SNL	**0.686**	−0.019	0.032
IPW	**0.684**	0.032	0.060
P2BL	**0.676**	−0.132	−0.091
IOW	**0.633**	**0.207**	**−0.313**
GW	**0.588**	0.034	−0.224
AL	**0.647**	−0.025	−0.167
D2L	**0.482**	−0.142	−0.337
QL	**0.498**	0.051	−0.106
P2L	**0.623**	−0.080	−0.131
CFL	**0.462**	0.039	**−0.280**
Canonical R	0.873	0.822	0.670

pectoral fins, and narrow interorbital distances. Midwater darters remain relatively fusiform (the characteristic of a body being broadest in the middle and tapering at both ends); benthic darters exhibit reduced fusiformity resulting mainly from varying combinations of a flattened venter, an anterior shift of the greatest body depth (riffle species), a deep caudal peduncle (riffle species), and a blunter snout (pool and some riffle species). Living on the bottom of a stream in either a riffle or a pool lessens exposure to current (Hynes 1970) and is accompanied by a reduction in streamlining.

Presumably, current flowing against the large pectoral fins of a riffle darter pushes down and helps the darter maintain its position on the substrate. Small paired fins on fishes reduce drag, especially when the fins are held against the body (Alexander 1967). Among darters, small pectoral fins are most characteristic of midwater species. Although the pectoral fin is large in benthic darters occupying both riffle and sluggish-water habitats, a previous relationship (third habitat and meristic canonical variates) indicated that the number of pectoral rays is higher only in riffle species. An obvious explanation is that more pectoral rays are needed for increased support in darters confronted by higher water velocity.

The third pair of canonical variates indicates that darters living over sand in deep water of large streams have shallow bodies, narrow caudal peduncles, narrow interorbital distances, and short second dorsal and caudal fins. The slender body and short fins probably facilitate the sand-burying behavior of *Ammocrypta* and *E. vitreum*. Interestingly, narrow interorbital width was associated with benthicity on both the second and third canonical variates, despite the otherwise distinctiveness of the habitats characterizing these variates. For benthic darters, most of the visual field is above, and consequently their eyes are more dorsally located. Much of the visual field for midwater darters is below, requiring a more lateral position of the eyes, and hence a greater interorbital distance.

Overview of habitat specializations

The strong correlations between combinations of morphological characteristics and habitat variables suggest that the habitat has been a prime influence in the evolution of darter morphology. Among those characteristics are degree of fusiformity, pigmentation, premaxillary protrusility, snout shape, and loss of the swimbladder; the last two features are habitat- related, but also are discussed above as feeding specializations.

Unrelated groups of darters occupying the same habitats often resemble one another in gross morphology more than they resemble close relatives; outstanding examples are the convergence of the sand-run inhabiting *E. vitreum* with species of *Ammocrypta,* and of *P. roanoka* with riffle-inhabiting species of *Etheostoma.*

Gravel-run darters are midwater, large, fusiform, have pointed snouts, a large interorbital width, short pectoral fins, a relatively large swimbladder, nonprotusile premaxillaries, a prominent black basicaudal spot, usually lack bright colors, often have a black midlateral stripe, seasonally moderate to weak sexual dimorphism, and high meristic counts. Riffle darters are benthic, usually moderate in size, lack swimbladders, have reduced fusiformity, deep caudal peduncles, long pectoral and second dorsal fins, nonprotrusile premaxillaries, a narrow interorbital width, are usually brightly pigmented, often have vertical bars and pronounced sexual dimorphism, and lack a prominent black basicaudal spot. Sand-run darters are benthic, moderate in size, translucent, have pointed snouts, protrusile premaxillaries, narrow caudal peduncles, are without swimbladders, have a weak black basicaudal spot and weak sexual dimorphism in coloration. Quiet-pool darters are benthic, small, lack swimbladders, have reduced fusiformity, blunt snouts, short second dorsal fins, a weak black basicaudal spot, usually have protrusile premaxillaries, and are usually dully pigmented with only moderate development of sexually dimorphic colors. Flowing-pool species may be midwater (e.g. *P. maculata* and closely related species) or benthic (e.g. most species of *Nanostoma*). If midwater, they share most characteristics of gravel-run darters but have a stronger tendency toward a narrow caudal peduncle; if benthic, they are similar to riffle darters but have a reduced premaxillary frenum, and lack notably deep caudal peduncles, long pectoral and dorsal fins, and dark vertical bars.

Reproductive specializations

One of the most interesting consequences of darters' occupation of a diversity of habitats has been the evolution of a diversity of reproductive strategies. Data on spawning, which vary from knowing only where eggs are laid to having films of spawning sequences, are available for seven species of *Percina* and 54 species of *Etheostoma*; nothing is known of the spawning habitats of *Ammocrypta*. Among the 61 species for which spawning behavior is known, four types of behavior exist (Table 5).

The most primitive type of spawning behavior, egg-burying, is characteristic of all seven species of *Percina* for which data are available (and presumably of all species of *Percina*) and of some species of *Etheostoma*. During egg-burying, the female works her body partially below the surface of the substrate and, with her genital papilla buried and a male mounted on her back, expels eggs (Winn 1958, Fig. 4). The substrates usually utilized are loose gravel, sand, or mixed gravel and sand. Eggs are abandoned and receive no parental care.

Egg-attaching, the second type of spawning behavior, is a derived behavior known only in *Etheostoma* and has arisen independently in several unrelated subgenera. The presence of both egg-burying and egg-attaching species within the subgenera *Etheostoma, Ozarka* and *Oligocephalus* (Table 5) suggests that egg-attaching is derived directly from egg-burying. The female selects the site of egg deposition (presumably within a male's territory in some species), typically a plant or rock, and, with the male following her, elevates to the site. As she does so, the male follows and mounts, the two vibrate, and eggs and sperm are released (Winn 1958, Fig. 4, Page et al. 1982, Fig. 1). From one to three adhesive eggs normally are released during each mating and are pushed by the female onto a plant, rock or other object. Attaching eggs to plants seems to involve behavior identical to that of attaching eggs to rocks; the substrate utilized probably is determined by what is readily available in the habitat more than anything else. Eggs are abandoned and receive no parental care.

The behaviors associated with egg-clumping and egg-clustering, the third and fourth types of spawning behavior in darters, are the most complex. Known only in some species of the subgenus *Nothonotus,* egg-clumping is almost certainly a derivative of egg-burying (Page et al. 1982). Egg-clustering, a refinement of egg-attaching behavior, is limited to the subgenera *Boleosoma* and *Catonotus*.

In both egg-clumping and egg-clustering, males select cavities under large rocks as territories and future nesting sites. Soon after occupying a cavity,

Table 5. Species in each of four categories of darter spawning behavior.

Buriers	Attachers	Clumpers	Clusterers
Percina (Alvordius) maculata	*E. (Etheostoma) blennioides*	*E. (Nothonotus) maculatum*	*E. (Boleosoma) nigrum*
P. (A.) peltata	*E. (Nanostoma) zonale*	*E. (N.) microlepidum*	*E. (B.) olmstedi*
P. (A.) notogramma	*E. (N.) coosae*	*E. (N.) aquali*	*E. (B.) perlongum*
P. (Ericosma) evides	*E. (N.) simoterum*		*E. (B.) longimanum*
P. (Hypohomus) aurantiaca	*E. (N.)duryi*		*E. (Catonotus) olivaceum*
P. (Cottogaster) copelandi	*E. (N.) barrenense*		*E. (C.) squamiceps*
P. (Percina) caprodes	*E. (N.) rafinesquei*		*E. (C.) neopterum*
Etheostoma (Litocara) nianguae	*E. (N.)* sp. (Red Snubnose)		*E. (C.) kennicotti*
E. (Etheostoma) tetrazonum	*E. (N.)* sp. (Lowland Snubnose)		*E. (C.) flabellare*
E. (E.) variatum	*E. (Ioa) vitreum*		*E. (C.) virgatum*
E. (Doration) stigmaeum	*E. (Vaillantia) chlorosomum*		*E. (C.) obeyense*
E. (Nothonotus) rufilineatum	*E. (Belophlox) okaloosae*		*E. (C.) barbouri*
E. (N.) camurum	*E. (Villora) edwini*		*E. (C.) smithi*
E. (N.) tippecanoe	*E. (Osarka) boschungi*		*E. (C.) striatulum*
E. (Fuscatelum) parvipinne	*E. (O.) trisella*		
E. (Ozarka) cragini	*E. (Oligocephalus) lepidum*		
E. (Oligocephalus) spectabile	*E. (O.) asprigene*		
E. (O.) caeruleum	*E. (O.) grahami*		
E. (O.) radiosum	*E. (O.) ditrema*		
	E. (Boleichthys) exile		
	E. (B.) fusiforme		
	E. (B.) gracile		
	E. (B.) proeliare		
	E. (B.) fonticola		
	E. (B.) microperca		

the male cleans it of silt and debris by vigorously shaking his tail and forcing out fine-particle materials. Eggs are amassed within the cavity by one or more females and guarded by the male until hatching.

Egg-clumping was first reported in 1939 for *E. maculatum* (Raney & Lachner 1939) but was unreported for other darters until observed in 1981 for *E. aquali* and *E. microlepidum* (Page et al. 1982). In egg-clumping the female swims into the cavity being guarded by the male, wedges herself into the interface between the stone and the gravel substrate beneath the stone, and deposits eggs. The resulting clump then is guarded by the male. Eggs are adhesive and adhere to both the nest stone and the underlying substrate material; if a stone is lifted from the water, a clump of adhesive eggs remains attached to it (Raney & Lachner 1939, Fig. 1, Page et al. 1982, Fig. 1). The transition from burying in some species of *Nothonotus* (Mount 1959, Stiles 1972, Trautman 1981) to clumping in others requires only that the male establish a territory beneath a stone and that a female wedge herself between the stone and the substrate; in a way, the eggs are still being buried.

In egg-clustering, eggs are arranged in a single-layer cluster on the underside of a stone earlier selected and guarded by the male. A female joins the male in his cavity under the stone, rolls over and simultaneously rises to press her belly against the underside of the stone. As she does so, the male also inverts, rises, and positions himself next to the female. Eggs and sperm are released, and a single layer cluster of eggs is formed on the underside of the stone (Page 1974, Figs. 6, 7). In some species the male and female intermittently and briefly invert to deposit eggs; in others they remain inverted for the duration of egg-laying (Winn 1958, Page 1975). When a female is spent, she leaves the cavity. Other females mate sequentially with the male. Clusters of eggs are kept clean by the guarding male. In typical ambient conditions, hatching takes 5–10 days (Page 1983).

Although it is remarkable that two non-sister groups of darters, *Boleosoma* and *Catonotus,* have evolved egg-clustering, it is even more remarkable that egg-clustering has evolved also in a group of minnows. The minnow genus *Pimephales,* with four species, also forms a single-layer cluster of eggs on the underside of a rock which is guarded by the male (McMillan & Smith 1974). Several morphological characteristics associated with egg-clustering in darters (discussed below) also are found in *Pimephales.*

The reproductive guilds recognized for percids by Balon et al. (1977) and Balon (1981) are similar in concept to the four categories of spawning behavior discussed above in that they rely heavily on the site of egg deposition. Egg-burying darters are referred to as brood hiding lithophils (rock and gravel spawners), attachers are open substrate phytophils or lithophils (plant or rock spawners), and clumpers and clusterers are nest spawning speleophils (cave spawners). The classification is useful because it places darters in guilds originally developed for all fishes (Balon 1975); unfortunately, several errors were made in classifying darters. *P. caprodes* is classified as an open substrate sand spawner, but usually spawns in gravel (Trautman 1948, Cross 1967) and buries its eggs (Winn 1958, Fig. 4) as do *P. copelandi, P. maculata,* and *E. caeruleum,* all classified by Balon et al. (1977) as brood hiding lithophils. *P. shumardi* is classified as a brood hiding lithophil, but we are unaware of any published documentation for that classification. *E. blennioides* is a phytophil (Fahy 1954, Winn 1958) but does not construct a nest or guard its eggs (although some indirect care may result from male territorial behavior). *E. blennioides* should be classified as an open substrate spawner.

Although it is true that species of *Ammocrypta* always are found in sand runs, no observations on their spawning habits have been made (Williams 1975), and it is premature to assign them to a reproductive guild. It is reasonable to call them psammophils (sand spawners), but it is more likely that they bury eggs rather than scatter them openly as Balon et al. (1977) indicated by classifying them as open substrate spawners.

The only published observation on reproduction in *E. vitreum* is that of Winn & Picciolo (1960) which does not state directly that eggs are hidden under stones and guarded. Although we have categorized *E. vitreum* as an attacher (Table 5), the behavior appears to be unusual and more observations are needed. As documentation for their classification of *E. vitreum* as a nest spawning speleophil, Balon et al. (1977) cite a publication (Page 1975) which does not mention *E. vitreum.*

Morphological specializations found in more than one subgenus and having distributions correlated with spawning behavior are 'modified' (enlarged and strongly toothed) scales, breeding tubercles, fin knobs, swollen flesh, and tubular versus flattened female genital papillae (Table 6).

Breeding tubercles and modified scales generally are restricted to males, reach their maximum development during the breeding season, and are used at least in part to provide tactile stimulation to the female during mating (Collette 1965, Wiley &

Table 6. Distribution of sexually selected morphological specializations among darters.

	Buriers	Attachers	Clumpers	Clusterers
Modified scales	X			
Breeding tubercles	X	X		
Swollen flesh				X
Fin knobs				X
Genital papillae				
Tubular	all	some		
Flat			all	all

Collette 1970, Page 1976). Among darters, they are restricted to egg-buriers and egg-attachers (Table 6), both of which move over a large area when spawning. Among clumpers and clusterers, the continuous close proximity of the male and nest stone and, except during the first mating act of the first female, the presence of eggs (see below) may negate the necessity for additional (i.e. tactile) stimulation of the female. Modified scales are found only on *Percina,* all of which probably are egg-buriers.

Although the flesh on the head and nape of the male in many species of darters swells slightly during the spawning season, it becomes extremely

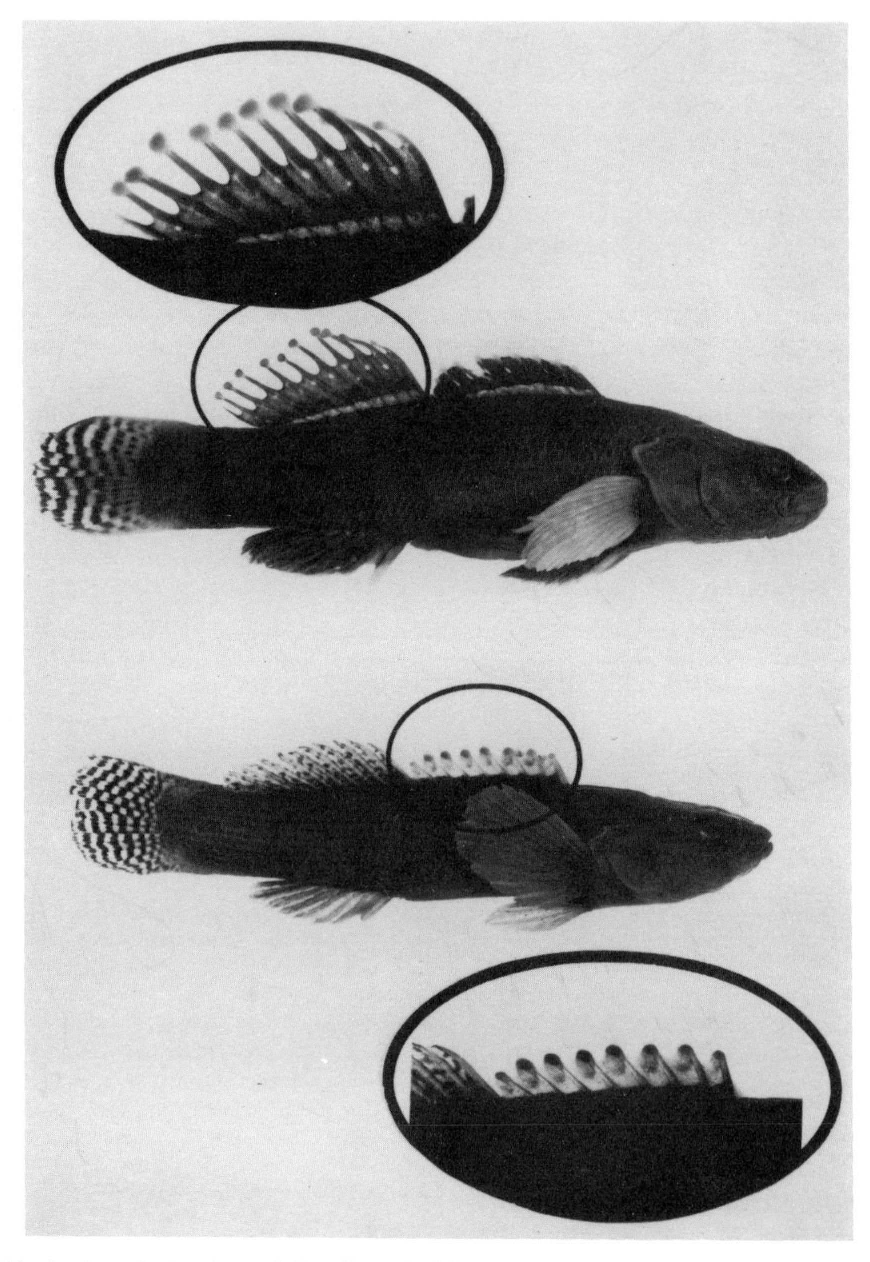

Fig. 8. Egg-mimicking knobs on the dorsal rays of a breeding male *Etheostoma neopterum* (top) and dorsal spines of a breeding male *E. flabellare* (bottom).

swollen in egg-clustering subgenera. The swollen flesh may have a secretory (i.e., fungicidal or bactericidal) function (Cross 1967, Smith 1979), or sensory and secretory functions similar to those described for swollen flesh on fathead minnow (*Pimephales promelas*) breeding males (Smith & Murphy 1974). If so, the function of the single-layer arrangement of eggs in egg-clustering darters and minnows is, at least in part, to permit the male to rub his head and nape over each egg. The multilayer arrangement of eggs in egg-clumping darters precludes selection for such care by the male.

Knobs develop on the tips of dorsal spines and rays of breeding males in several groups of egg-clustering darters. They are present on dorsal spines of *E. flabellare* and *E. kennicotti,* on dorsal spines and sometimes rays of the *E. squamiceps* group (*E. squamiceps, E. neopterum, E. olivaceum*), and sometimes are present on the dorsal spines of large *E. nigrum* and *E. olmstedi.* Knobs probably originated as fleshy masses on the tips of the dorsal spines in association with their ability to reduce the likelihood of rupturing eggs during nest guarding. The presence on dorsal rays perhaps originally was a pleiotropic effect of their presence on dorsal spines. However, the extreme development as large yellow knobs on the dorsal spines of male *E. flabellare* and *E. kennicotti,* and as large white to yellow knobs on the dorsal rays of male *E. neopterum* (Fig. 8) seems secondarily to have taken on a new function: egg-mimicry. If it is true that females are more likely to add eggs to an existing nest than to begin a new nest (Page 1974), then the selective advantage of egg-mimicking knobs is readily apparent. The knobs on the tips of the dorsal fin elements of a breeding male displaying to a female under a nest stone could easily appear to be eggs on the stone; in fact, because of the small area under the stone, the knobs probably are pressed against the stone as the male displays to the female. Similar egg-mimicry (as egg-like yellow spots on the anal fin of the male) stimulates fin nibbling during spawning in females of mouth-brooding cichlids (e.g. *Haplochromis* spp.) (Wickler 1962, Keenleyside 1979).

Specializations in reproductive behavior have been accompanied by modifications of the genital papilla of the female and, although the papillae are highly variable, in many instances it is possible to relate the structure of the papilla to the reproductive behavior (Page 1983). The papillae of egg-buriers are conical tubes which inject the eggs into the substrate. In egg-clumping and egg-clustering species, the female has a wide, flat papilla which enables her to push an egg, as it is released from the papilla, against the underside of a stone. In egg-attaching species a diverse array of papillae exists, but many are tubular.

E. sellare is a highly unusual darter (Knapp 1976, Page 1981) and, unfortunately, is on the verge of extinction. Nothing is known of its spawning behavior, but one of its unusual characteristics is that the female possesses a flattened genital papilla (Knapp 1976). If *E. sellare* is an egg-clumping or egg-clustering species, one of the important management procedures will be to insure the presence of adequate clumping or clustering sites.

Acknowledgements

We wish to thank C.W. Ronto for preparation of the illustrations, and K.D. Fausch, J.R. Karr, M.E. Retzer, R.B. Selander, P.W. Smith, M.J. Wiley, and two anonymous reviewers for constructive comments on the manuscript.

References cited

Alexander, R.M. 1967. Functional design in fishes. Hutchinson University Library, London. 160 pp.

Bailey, R.M & W.A. Gosline. 1955. Variation and systematic significance of vertebral counts in the American fishes of the family Percidae. Misc. Publ. Mus. Zool. Univ. Mich. 93: 1–44.

Baker, J.A. & S.T. Ross. 1981. Spatial and temporal resource utilization by southeastern cyprinids. Copeia 1981: 178–189.

Balon, E.K. 1975. Reproductive guilds of fishes: a proposal and definition. J. Fish. Res. Board Can. 32: 821–864.

Balon, E.K. 1981. Additions and amendments to the classification of reproductive styles in fishes. Env. Biol. Fish. 6: 377–389.

Balon, E.K., W.T. Momot & H.A. Regier. 1977. Reproductive guilds of percids: results of the paleogeographical history and ecological succession. J. Fish. Res. Board Can. 34: 1910–1921.

Barlow, G.W. 1972. The attitude of fish eye-lines in relation to body shape and to stripes and bars. Copeia 1972: 4–12.

Bruun, A.F. 1940. A study of a collection of the fish *Schindleria* from south Pacific Waters. Dana. Rep. 21: 1–12.

Burr, B.M. & L.M. Page. 1978. The life history of the cypress darter, *Etheostoma proeliare,* in Max Creek, Illinois. Ill. Nat. Hist. Surv. Biol. Notes 106: 1–15.

Burr, B.M. & L.M. Page. 1979. The life history of the least darter, *Etheostoma microperca,* in the Iroquois River, Illinois. Ill. Nat. Hist. Surv. Biol. Notes 112: 1–15.

Collette, B.B. 1962. The swamp darters of the subgenus *Hololepis* (Pisces, Percidae). Tulane Stud. Zool. 9: 115–211.

Collette, B.B. 1965. Systematic significance of breeding tubercles in fishes of the family Percidae. Proc. U.S. Nat. Mus. 117: 567–614.

Cooley, W.W. & P.R. Lohnes. 1971. Multivariate data analysis. Wiley, New York. 364 pp.

Cross, F.B. 1967. Handbook of fishes of Kansas. Misc. Publ. Mus. Nat. Hist. Univ. Kan. 45: 1–357.

Distler, D.A. 1972. Observations on the reproductive habits of captive *Etheostoma cragini* Gilbert. Southwest. Nat. 16: 439–441.

Dixon, W.J., M.B. Brown, L. Engelman, J.W. Frane, M.A. Hill, R.I. Jennrich & J.D. Toporek. 1981. BMDP statistical software. University of California Press, Berkeley. 725 pp.

Estes, R.D. 1974. Social organization of the African Bovidae. pp. 106–205. *In*: V. Geist & F. Walther (ed.) Behavior of Ungulates and its Relation to Management, IUCN, Morges.

Fahy, W.E. 1954. The life history of the northern greenside darter, *Etheostoma blennioides blennioides* Rafinesque. J. Elisha Mitchell Sci. Soc. 70: 139–205.

Forbes, S.A. & R.E. Richardson. 1908. The fishes of Illinois. Illinois Natural History Survey, Urbana. 357 pp.

Gould, S.J. & R.C. Lewontin. 1979. The spandrels of San Marco and the Panglossian paradigm: a critique of the adaptationist programme. Proc. R. Soc. London Biol. Sci. 205: 581–598.

Harris, R.J. 1975. A primer of multivariate statistics. Wiley, New York. 332 pp.

Hemmingsen, A.M. 1934. A statistical analysis of the differences in body size of related species. Videnskabelige Meddelselsen Dansk Naturhistorisk Farening Kobenhaven 98: 125–160.

Hynes, H.B.N. 1970. The ecology of running waters. University of Toronto Press, Toronto. 555 pp.

Jaksic, F.M. 1981. Recognition of morphological adaptations in animals: the hypothetico-deductive method. BioScience 31: 667–670.

Jordan, D.S. & B.W. Evermann. 1896. The fishes of North and Middle America: a descriptive catalog of the species of fish-like vertebrates found in the waters of North America, north of the Isthmus of Panama. U.S. Natl. Mus. Bull. 47: 1–1240.

Keast, A. & D. Webb. 1966. Mouth and body form relative to feeding ecology in the fish fauna of a small lake, Lake Opinicon, Ontario. J. Fish. Res. Board Can. 23: 1845–1874.

Keenleyside, M.H.A. 1979. Diversity and adaptation in fish behavior. Springer-Verlag, New York. 208 pp.

Kerr, S.R. 1971. Prediction of fish growth efficiency in nature. J. Fish. Res. Board Can. 28: 809–814.

Kirchner, T.B., R.V. Anderson & R.E. Ingham. 1980. Natural selection and the distribution of nematode sizes. Ecology 61: 232–237.

Knapp, L.W. 1976. Redescription, relationships and status of the Maryland darter, *Etheostoma sellare* (Radcliffe and Welsh), an endangered species. Proc. Biol. Soc. Wash. 89: 99–118.

Kuehne, R.A. & J.W. Small, Jr. 1971. *Etheostoma barbouri,* a new darter (Percidae, Etheostomatini) from the Green River with notes on the subgenus *Catonotus.* Copeia 1971: 18–26.

Lee, D.S. & R.E. Ashton, Jr. 1979. Seasonal and daily activity patterns of the glassy darter, *Etheostoma vitreum* (Percidae). ASB (Assoc. Southeast. Biol.) Bull. 26: 36.

Lee, D.S., C.R. Gilbert, C.H. Hocutt, R.E. Jenkins, D.E. McAllister & J.R. Stauffer, Jr. 1980. Atlas of North American freshwater fishes. N.C. State Mus. Nat. Hist., Raleigh. 867 pp.

Lowenstein, O. 1957. The sense organs: the acousticolateralis system. pp. 155–186. *In*: M.E. Brown (ed.) The Physiology of Fishes, Academic Press, New York.

McMillan, V.E. & R.J.F. Smith. 1974. Agonistic and reproductive behavior of the fathead minnow (*Pimephales promelas* Rafinesque). Z. Tierpsychol. 34: 25–58.

Mendelson, J. 1975. Feeding relationships among species of *Notropis* (Pisces: Cyprinidae) in a Wisconsin stream. Ecol. Monogr. 45: 199–230.

Miller, P.J. 1979. Adaptiveness and implications of small size in teleosts. pp. 163–306. *In*: P.J. Miller (ed.) Fish Phenology: Anabolic Adaptiveness in Teleosts, Academic Press, London.

Mount, D.I. 1959. Spawning behavior of the bluebreast darter, *Etheostoma camurum* (Cope). Copeia 1959: 240–243.

Page, L.M. 1974. The life history of the spottail darter, *Etheostoma squamiceps,* in Big Creek, Illinois, and Ferguson Creek, Kentucky. Ill. Nat. Hist. Surv. Biol. Notes 89: 1–20.

Page, L.M. 1975. Relations among the darters of the subgenus *Catonotus* of *Etheostoma.* Copeia 1975: 782–784.

Page, L.M. 1976. The modified midventral scales of *Percina* (Osteichthyes; Percidae). J. Morph. 148: 255–264.

Page, L.M. 1977. The lateralis system of darters (Etheostomatini). Copeia 1977: 472–475.

Page, L.M. 1981. The genera and subgenera of darters (Percidae, Etheostomatini). Occ. Pap. Mus. Nat. Hist. Univ. Kansas 90: 1–69.

Page, L.M. 1983. The handbook of darters. T.F.H. Publications, Neptune City. 271 pp.

Page, L.M., M.E. Retzer & R.A. Stiles. 1982. Spawning behavior in seven species of darters (Pisces: Percidae). Brimleyana 8: 135–143.

Pflieger, W.L. 1978. Distribution, status, and life history of the Niangua darter, *Etheostoma nianguae.* Mo. Dept. Conserv. Aquatic Ser. 16: 1–25.

Pimentel, R.A. 1979. Morphometrics: the multivariate analysis of biological data. Kendall/Hunt Publ. Co., Dubuque. 276 pp.

Raney, E.C. & E.A. Lachner. 1939. Observations on the life history of the spottail darter, *Poecilichthys maculatus* (Kirtland). Copeia 1939: 157–165.

Ricker, W.E. 1973. Linear regressions in fishery research. J. Fish. Res. Board Can. 30: 409–434.

Schenck, J.R. & B.G. Whiteside. 1977. Food habits and feeding behavior of the fountain darter, *Etheostoma fonticola* (Osteichthyes: Percidae). Southwest. Nat. 21: 487–492.

Schoener, T.W. 1971. Theory of feeding strategies. Ann. Rev. Ecol. Syst. 2: 369–404.

Smith, R.J.F. 1979. Alarm reaction of Iowa and johnny darters (*Etheostoma,* Percidae, Pisces) to chemicals from injured conspecifics. Can. J. Zool. 57: 1278–1282.

Smith, R.J.F. & B.D. Murphy. 1974. Functional morphology of the dorsal pad in fathead minnows (*Pimephales promelas* Rafinesque). Trans. Amer. Fish. Soc. 103: 65–72.

Stiles, R.A. 1972. The comparative ecology of three species of *Nothonotus* (Percidae-*Etheostoma*) in Tennessee's Little River. Ph.D. Thesis, University of Tennessee, Knoxville. 97 pp.

Strawn, K. 1956. A method of breeding and raising three Texas darters. Part II. Aquarium J. 27: 12–14, 17, 31–32.

Surat, E.M., W.J. Matthews & R.J. Bek. 1982. Comparative ecology of *Notropis albeolus, N. ardens* and *N. cerasinus* (Cyprinidae) in the upper Roanoke River drainage, Virginia. Amer. Midl. Nat. 107: 13–24.

Thomas, D.L. 1970. An ecological study of four darters of the genus *Percina* (Percidae) in the Kaskaskia River, Illinois. Ill. Nat. Hist. Surv. Biol. Notes 70: 1–18.

Trautman, M.B. 1948. A natural hybrid catfish, *Schilbeodes miurus* x *Schilbeodes mollis.* Copeia 1948: 166–174.

Trautman, M.B. 1981. The fishes of Ohio. Ohio State University Press, Columbus. 782 pp.

Wehnes, R.A. 1973. The food and feeding interrelationships of five sympatric darter species (Pisces: Percidae) in Salt Creek, Hocking County, Ohio. M.Sc. Thesis, Ohio State University, Columbus. 79 pp.

Wickler, W. 1962. 'Egg dummies' as natural releasers in mouthbreeding cichlids. Nature, Lond. 194: 1092–1093.

Wiley, M.L. & B.B. Collette. 1970. Breeding tubercles and contact organs in fishes: their occurrence, structure, and significance. Bull. Amer. Mus. Nat. Hist. 143: 143–216.

Williams, J.D. 1975. Systematics of the percid fishes of the subgenus *Ammocrypta,* genus *Ammocrypta,* with descriptions of two new species. Bull. Ala. Mus. Nat. Hist. 1: 1–56.

Winn, H.E. 1958. Comparative reproductive behavior and ecology of fourteen species of darters (Pisces-Percidae). Ecol. Monogr. 18: 155–191.

Winn, H.E. & A.R. Picciolo. 1960. Communal spawning of the glassy darter *Etheostoma vitreum* (Cope). Copeia 1960: 186–192.

Originally published in Env. Biol. Fish. 11: 139–159

A portable camera box for photographing small fishes

Lawrence M. Page & Kevin S. Cummings
Illinois Natural History Survey, Champaign, IL 61820, U.S.A.

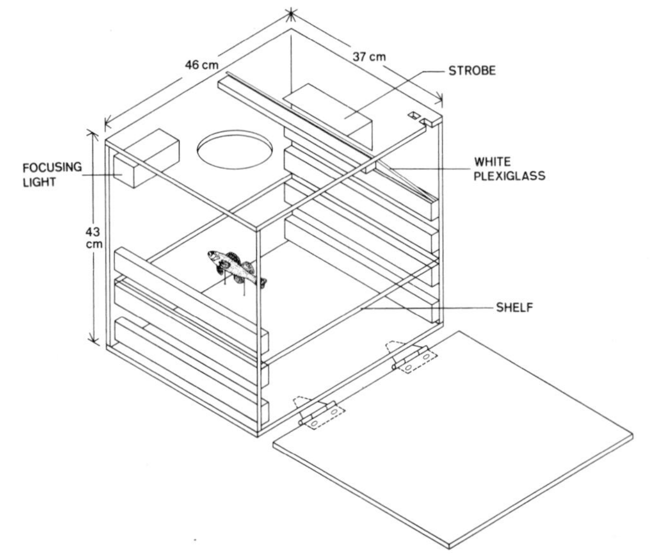

Fishes are difficult to photograph for a variety of reasons. Ideally, they should be photographed alive and in their natural environments. However, few of us can manage the time and aquatic skills necessary to accomplish that. Also, many species live in turbid water or at insufficient depths to accommodate underwater photography. Live fishes in aquary often are frightened into only a pale reminder of their former selves, and settings seldom simulate natural conditions. The third alternative is to preserve the fishes and photograph them. With proper controls good life-like images can result. The apparatus described below works well for small fishes and is portable. Photos in the 'Handbook of Darters' (T.F.H. Publications, Inc., Neptune City, NJ 07753) were taken with this procedure.

Fishes are preserved in strong (≈50%) formalin, and usually die with median fins erect. If it appears that a fish is not going to die in a photogenic state, its fins and jaws can be manipulated once the fish is dead but not yet 'fixed'. There is a loss of brightness; the iridescence of live fishes is lost almost immediately. But most colors remain for several hours, and the photograph will come close to expressing life colors of the fish. Once fixed, the fish is impaled on two pins embedded in a clear plexiglass shelf (6 mm thick) temporarily removed from the camera box.

The box, constructed of 10 mm plywood, has evolved through a series of prototypes. The one shown above, with three shelf levels, will accommodate specimens up to about 15 cm in length. The top level is used for the smallest specimens which, in order to fill the frame, need to be near the camera. Along one upper edge of the box is a piece of white plexiglass (3 mm thick), angled at about 45°. A strobe is placed behind the plexiglass and aimed at the fish. The cord from the strobe to the camera passes through a hole in the box. Along another edge of the box is a focusing light. An AC cord from both the strobe and focusing light goes through a hole in the box to an electric outlet. The inside of the box is painted flat black, except for the side opposite the strobe and the bottom, which are painted white. Other colors can be used as background, but most result in a reflection of the camera lens; white and gray backgrounds work best.

The shelf and impinged fish are placed in the box so that the light strikes the fish dorsally. The lens of the camera is placed through an opening in the top and focused on the fish. The door is shut, and the fish is photographed. In present use is a 35-mm camera with close-up lenses and a Vivitar 252 strobe with an AC adapter. This strobe is a perfect size, but unfortunately it is no longer made. A stronger light requires additional opaqueness between it and the fish. We use Ektachrome 64 film, and shoot most fish at F11 or F11.5.

David G. Lindquist & Lawrence M. Page (ed.), Environmental biology of darters. ISBN 90-6193-506-7

Species and subject index

Annelida 79
Age 45, 61, 71, 83
Aggression 52, 76, 101
Alvordius 106, 111
Ammocrypta 103, 104, 105, 106, 107, 108, 113, 114, 116, 117, 119
  asprella 109
  beani 47, 61, 109
  bifascia 63
  clara 109
  pellucida 108
  spp. 68
Appalachians 9
Arachnida 49, 79
Artemia salina 24
Benthicity 24, 103
Benthos 103, 105, 108
Body size 31, 103
Boleichthys 75, 104, 106, 113, 114
Boleosoma 115, 117, 119
Breeding
  behavior 31, 52, 103
  habitat 31, 53, 103
  season 31, 52, 61, 71, 85
Campostoma anomalum 96
Canonical correlation 105
Carpiodes carpio 73
Catfishes 31
Catonotus 75, 106, 113, 117, 119
Cheumatopsyche spp. 48
Coloration 50, 73, 110, 114
  cryptic 106, 108, 113
Competition 10, 18, 37, 48, 55, 101, 103
Critical thermal maxima 95
Crustacea 40, 42, 48, 79, 105
Cyprinodon elegans 101
Demography 50, 73
Density 73, 88
Development 21, 76, 86
Diet 37, 48, 59, 71, 103
Dispersal 9, 21
Echinorhyncus species 78
Eggs 22, 31, 43, 53, 61, 85
Electrophoresis 9
Eleocharis spp. 47
Ericosma 83
Ericymba buccata 96
Erimyzon oblongus 73
Etheostoma 28, 29, 45, 48, 54, 88, 103, 104, 105, 106, 107, 108, 116, 117
  aquali 109, 118
  asprigene 45, 71, 73, 74, 75, 118
  australe 74, 109
  barbouri 43, 50, 118
  barrenense 118
  blennioides 31, 54, 55, 73, 76, 96, 108, 114, 118, 119
  blennius 109, 115
  boschungi 68, 118
  caeruleum 21, 28, 31, 37, 55, 71, 96, 108, 118, 119
  camurum 118
  chlorosomum 111, 113, 115, 118
  cinereum 109
  coosae 68, 118
  cragini 113, 118
  ditrema 45, 73, 74, 75, 118
  duryi 118
  edwini 118
  euzonum 9
  erizonum 9
  exile 28, 115, 118
  flabellare 21, 28, 37, 52, 54, 73, 95, 108, 109, 118, 120, 121
  fonticola 28, 68, 105, 111, 113, 115, 118
  fusiforme 63, 115, 118
  gracile 115, 118
  grahami 74, 115, 118
  histrio 63
  hopkinsi 45
  jordani 85
  kanawhae 9
  kennicotti 50, 54, 111, 113, 118, 121
  lepidum 28, 68, 71, 74, 75, 115, 118
  longimanum 73, 115, 118
  luteovinctum 109
  maculatum 115, 118
  mariae 109
  microlepidum 115, 118
  microperca 42, 73, 105, 111, 113, 115, 118
  (*Nanostoma*) sp. 85
  neopterum 118, 120, 121
  nianguae 108, 118
  nigrum 27, 28, 31, 35, 37, 95, 108, 114, 115, 118, 121
  nuchale 45
  obeyense 118
  okaloosae 118
  olivaceum 73, 118, 121
  olmstedi 31, 34, 54, 73, 115, 118, 121
  osburni 9
  parvipinne 118
  perlongum 28, 31, 68, 115, 118
  proeliare 68, 73, 105, 111, 113, 115, 118
  radiosum 68, 71, 73, 74, 75, 76, 101, 118
    cyanorum 43
  rafinesquei 118
  rufilineatum 118
  sagitta 108, 109
  sellare 109, 115, 121
  simoterum 73, 118
  smithi 50, 54, 73, 118
  spectabile 28, 31, 37, 55, 68, 71, 74, 75, 76, 96, 101, 118

squamiceps 28, 54, 105, 108, 109, 118, 121
stigmaeum 47, 63, 85, 118
striatulum 73, 118
swaini 45, 63, 73, 74
tetrazonum 9, 55, 115, 118
tippecanoe 106, 118
trisella 118
tuscumbia 109
variatum 9, 73, 115, 118
virgatum 118
vitreum 106, 108, 116, 118, 119
whipplei 72
zonale 55, 73, 96, 108, 114, 118
Evolution 9, 21, 103
constraints 103
Experimental spawning cover 31
Fecundity 47, 53, 61, 71, 83
Feeding 37, 45, 71, 83, 103
first 21
periodicity 88
selective 83
Fundulus olivaceus 73
Gambusia
affinis 73
nobilis 101
Gastropoda 40, 49
Gene flow 21
Gonadosomatic index 52, 67, 74, 85
Growth 45, 61, 71, 83
Habitat 21, 31, 37, 45, 55, 61, 71, 83, 95, 103
Hadropterus 111
Haplochromis 121
Hybopsis lineapunctata 85
Hydroptila spp. 48
Hypentelium
etowanum 85
nigricans 96
Ictalurus natalis 73
Jaccard coefficient of association 37
Ichthyomyzon gagei 47
Imostoma 85, 113
Insecta 21, 37, 40, 45, 48, 71, 77, 79, 83, 89, 91, 103, 105, 108
Lake Waccamaw 31
Length/weight relationship 50, 87
Lepomis 114
macrochirus 73, 78
megalotis 73, 78
Life history 45, 61, 71, 83
Longevity 50, 61, 71, 83
Lowland snubnose darter 109, 118
Madtoms 31
Maturity 61, 71, 73, 83
Mayaca fluviatilis
Micropterus
punctulatus 78, 85
salmoides 78, 85
Midwater 103
Minnows 40, 103, 108
Morphology 103
Nanostoma 108, 113, 117
Nest
choice 31
egg number 31
fidelity 31
quality 31
Nematoda 79
Niche overlap 56
Nocomis leptocephalus 73
Notemigonus crysoleucas 114
Nothonotus 104, 117
Notropis 101
callistius 85
chrysocephalus 72, 85
gibbsi 85
longirostris 63, 73
roseipinnis 48
stilbius 85
texanus 48
venustus 63, 72, 85
volucellus 73
Noturus
gyrinus 32, 34
leptacanthus 47, 85
sp. 32, 34
Odontopholis 16, 111
Oligocephalus 45, 71, 117
Oligochaeta 40, 49
Ontogeny 21
Orontium aquaticum 47
Ova 22, 31, 43, 53, 61
Ovarian weight 47, 61, 76
Ozarka 106, 113, 117
Ozarks 9, 55
Parasites 71
Pearson's correlation coefficient 37
Percina 28, 29, 83, 85, 86, 92, 103, 104, 105, 106, 107, 108, 111, 113, 117, 120
(*Alvordius*) sp. 85, 86
antesella 108
aurantiaca 118
caprodes 21, 28, 42, 87, 92, 105, 108, 109, 111, 118, 119
semifasciata 21
carbonaria 109, 111
copelandi 108, 115, 118, 119
cymatotaenia 16, 108, 109, 111, 115
evides 83, 108, 111, 118
macrocephala 28
macrolepida 68
maculata 28, 43, 87, 88, 92, 108, 111, 117, 118, 119
nasuta 109, 111
nigrofasciata 47, 63, 68, 90, 92, 109, 115
notogramma 118
ouachitae 72, 108
palmaris 83, 108
peltata 85, 86, 90, 118
phoxocephala 87, 88, 92, 108, 109, 111
roanoka 108, 109, 116
sciera 28, 88, 92, 108, 109, 111
shumardi 43, 87, 92, 108, 109, 111, 119
tanasi 28
uranidea 109
Pimephales 119
notalus 73
promelas 121
vigilax 73
Pleistocene glaciation 9
Podostemum 85, 88
Polychaetes 28
Predation 21, 37, 78, 103, 105, 106, 108
Principal component analysis 56
Pylodictis olivaris 85
Red snubnose darter 109, 118
Refugia 9
Relict populations 9
Reproduction 31, 45, 61, 71, 83, 103
guilds 21, 68, 75
success 31
Resource partitioning 21, 37, 55, 101, 104
Rhinichthys 40
atratulus 96
Schoener index 37, 85
Sculpins 40
Selection 21, 104
Sexual 103, 104, 106
Sex ratio 53, 61, 71
Sexual dimorphism 50, 73, 83, 103
Semotilus atromaculatus 73, 78, 96
Sparganium americanum 47, 52
Spawning
behavior 31, 52, 103
cover 31
habitat 31, 53, 103
season 31, 52, 61, 71, 83
Speciation 21
Spearman rank correlation coefficient 37
Stenonema 48
Stizostedion
lucioperca 27
vitreum 27
Stream size 37, 58
Substrate size 55
Survival 50, 67, 76, 83
Swainia 111

Sympatry 37, 55, 101
Temperature
  preference 95
  tolerance 95
Territoriality 76, 86, 101, 103, 106, 113, 117
Vaillantia 113
Variation
  genic 9
  geographic 9
Vegetation types 55
Vitelline circulation 21
Wagner trees 11
Water
  current 47, 55, 85
  depth 47, 55, 85
Yolk supply 21
Zoogeography 9